Exploring Scale Symmetry

Fractals and Dynamics in Mathematics, Science, and the Arts: Theory and Applications

ISSN: 2382-6320

Published

For the complete list of volumes in this series, please visit
www.worldscientific.com/series/fds

Fractals and Dynamics in Mathematics, Science, and the Arts: Vol. 6   Theory and Applications

Exploring Scale Symmetry

Thomas Lowe

World Scientific

NEW JERSEY · LONDON · SINGAPORE · BEIJING · SHANGHAI · HONG KONG · TAIPEI · CHENNAI · TOKYO

Published by

World Scientific Publishing Co. Pte. Ltd.

5 Toh Tuck Link, Singapore 596224

USA office: 27 Warren Street, Suite 401-402, Hackensack, NJ 07601

UK office: 57 Shelton Street, Covent Garden, London WC2H 9HE

Library of Congress Cataloging-in-Publication Data

Names: Lowe, Thomas, author.

Title: Exploring scale symmetry / Thomas Lowe.

Description: New Jersey : World Scientific, [2021] | Series: Fractals and dynamics in mathematics, science, and the arts: theory and applications, 2382-6320 ; vol. 6 | Includes bibliographical references and index.

Identifiers: LCCN 2020041564 (print) | LCCN 2020041565 (ebook) | ISBN 9789813278547 (hardcover) | ISBN 9789813278554 (ebook) | ISBN 9789813278561 (ebook other)

Subjects: LCSH: Symmetry. | Quantum theory.

Classification: LCC Q172.5.S95 L69 2021 (print) | LCC Q172.5.S95 (ebook) | DDC 516/.1--dc23

LC record available at https://lccn.loc.gov/2020041564

LC ebook record available at https://lccn.loc.gov/2020041565

British Library Cataloguing-in-Publication Data

A catalogue record for this book is available from the British Library.

For any available supplementary material, please visit
https://www.worldscientific.com/worldscibooks/10.1142/11219#t=suppl

Desk Editors: George Vasu/Kwong Lai Fun

Typeset by Stallion Press
Email: enquiries@stallionpress.com

For Monica, Maurice and Ollie.

Preface

This book is all about the wonderful structures that you can find in the field of scale-symmetric geometry. It is a visual exploration through the less charted regions of this landscape of shapes, searching for extreme and unusual specimens and describing their construction and why they are interesting. Most of all this is a visual book, with lots of images, and graphical descriptions favoured over equations. As such, I expect it to be accessible to mathematicians, programmers, artists and hobbyists alike, though a background in geometry and maths is certainly recommended.

I first began exploring scale-symmetric shapes 10 years ago with help from the online community at fractalforums,[1] in this period, I have investigated lots of geometric constructions as a hobby. Some artificial in appearance, some realistic, some animating and some interactive. The common theme is that they are all scale-symmetric, and it is these explorations and findings that I present here. So, this is not a reference book on the current findings in the field, it is instead a description of my own path through this wide and diverse landscape. It follows the process of my exploration of the field, where one idea leads to the next, seeking out only the new, the beautiful and the interesting. Because of this, the majority of examples will be novel and previously uncharacterised. This sits well with the goal of this book: for readers to see what is possible within the field and to use the examples as a stepping stone for their own ideas.

---

[1] www.fractalforums.org.

About the Author

Thomas Lowe works as an Experimental Scientist in the Robotics Group at CSIRO in Australia. He has pursued the topic of scale symmetry for many years, developing several well-known structures, such as the Mandelbox and pyramidal surface fractal. These formulae are popular in the field of fractal art and have been used by special effects studios, such as Luma Pictures, on science fiction and fantasy movies. He has an academic background in computer science and a professional history in 3D computer games and in physically based animation.

Contents

Chapter 1

Introduction

The field of geometry has been energised in recent decades by the study of *scale symmetry* with its intricate beauty and deep connection to the natural world. But what is *scale symmetry*, and how did it become so popular? Let's begin by defining the term as we use it throughout the book.

What is Scale Symmetry?

Visually, it is the symmetry that describes structures such as these:

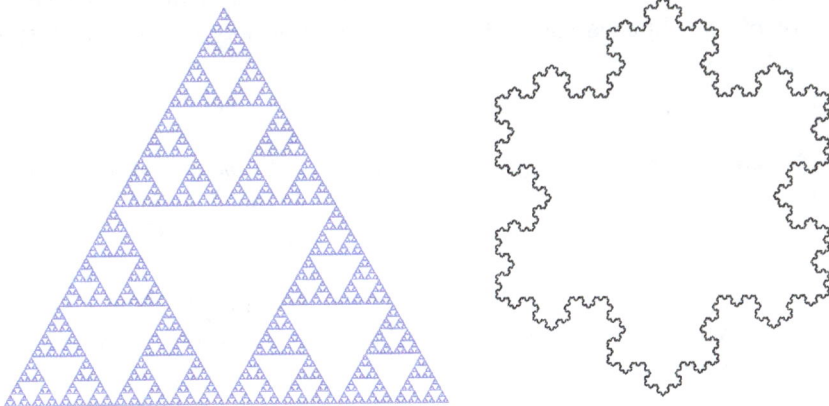

Figure 1.   Scale-symmetric structures. Above: *Sierpinski triangle* and *Koch snowflake*. Next page: *Mandelbrot set* and *Menger Sponge*.

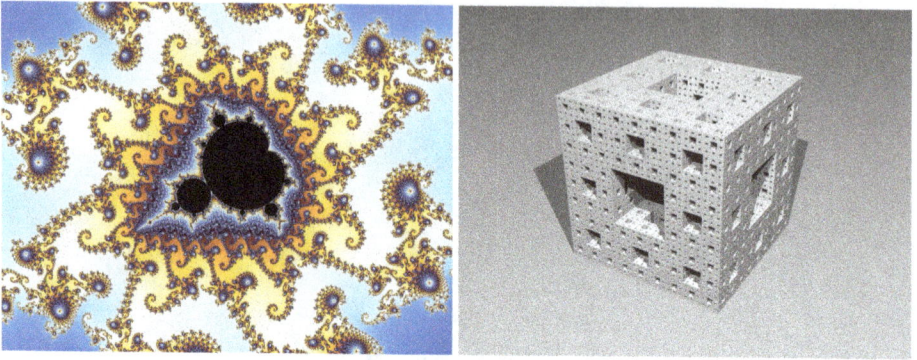

Figure 1.   (*Continued*)

To get to a definition, let's start by recalling the four elementary symmetries that we learnt at school:

Figure 2.   The four basic symmetries: translation, rotation, reflection and scale symmetry.

The last of these is scale symmetry. Scale-symmetric structures look the same under some sort of scaling operation. In the above example, this is a simple dilation by a factor of two, but we can include more general operations as well to give a broader definition:

Scale Symmetry: *The branch of geometry concerned with shapes that look the same under shape-preserving transformations that include scaling.*

Here, *shape-preserving* means transformations that preserve the shape of small features, this includes any combination of the four operations above: translation, rotation, reflection and scale. It is helpful to now qualify this broad definition with some specific types:

Pure scale-symmetric shapes are symmetric under *only* scaling, which results in radial structures. More general scale-symmetric shapes can include additional operations, in particular, rotation, which results in spiral patterns.

Figure 3. Left and middle: *pure scale symmetry* (continuous and discrete). Right: general *scale symmetry*, which includes other operations, in this case, rotation.

These symmetries generate radial and spiral designs, which are nice but rather limited. So, in this book, we are interested in the more diverse case of *multiple scale symmetry*, where shapes are symmetric under more than one transformation:

A Set with *Multiple Scale Symmetry: The shape within any region is identical to two or more smaller parts of the set under shape-preserving transformations.*

This definition includes fractals as a special case and many other recursive structures.

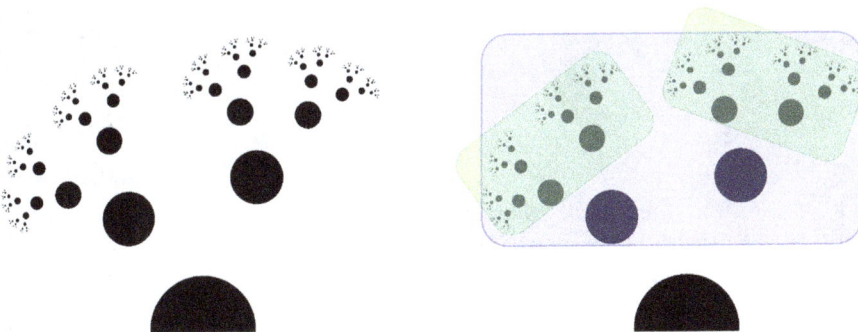

Figure 4. *Multiple scale symmetry*: Left shows a typical branching structure. Right shows an arbitrary bounding region (blue) and two regions (green) that are identical to the blue region under shape-preserving transformations.

These scale-symmetric objects are unbounded (infinitely large) by default, but we can also define *locally scale-symmetric* objects. These have their symmetry within only a finite scale range and finite extents. For example, a tree structure may be finite in size, and its recursive structure of branches may only go down to some minimum branch size.

Figure 5.   *Globally scale-symmetric* tree (left) and *locally scale-symmetric* tree with a maximum scale (middle) and an additional minimum scale (right).

The examples above are all *exactly scale-symmetric* as they are mathematically precise. There are also *statistically scale-symmetric* structures such as from stochastic algorithms like *Diffusion Limited Aggregation*, which show precise scale symmetry only on average. These in turn can be distinguished from *approximately scale-symmetric* structures, which are inexact, such as those found in nature like clouds and mountains. This qualifier *approximately* is however often omitted for natural objects, as it is always assumed to be the case.

Figure 6.   From left to right: *exactly*, *statistically* and *approximately* scale-symmetric structures on a *Vicsek fractal*, *Diffusion Limited Aggregation* and a frosted spider's web, respectively.

So, these are the basic types of scale symmetry, and they encompass a wide range of structures. Not just mathematical constructions but also natural examples, such as the trees, mountains, clouds and water. They all share the property that small parts of the object look similar to larger parts, they are self-similar:

Figure 7.   Examples of natural scale-symmetric structures.

Self-similar is just another term for *scale-symmetric*, so these two terms can be used interchangeably. In Chapter 3, we consider a definition of *self-similar* that is slightly different and more restrictive. But in general, *self-similar* and *scale-symmetric* can be thought of as equivalent terms.

Another natural example of scale symmetry is lightning, where the small arcs have a shape similar to the larger ones, and even within an arc, smaller sections look similar to larger sections, giving a rough appearance similar to fractures, cracks and maps of rivers.

Figure 8.   Lightning is scale-symmetric in its branching pattern and also in the shape of each arc.

As can be seen, scale-symmetric objects are particularly effective at describing the natural world. This was somewhat of a revelation when it was realised last century. After all, the geometry of objects that were symmetric under translation, reflection and rotations had been studied since the time of Plato and Euclid, and the subject had solidified into a precise and artificial-looking description of lines, planes, polygons, *Plato*nic solids (such as cubes and octahedrons) and other smooth forms, now called *Euclid*ean geometry. But the fourth and final shape-preserving transformation — scale — was not extensively studied until the last century. So, it came as a surprise when researchers such as Helge Von Koch, Michael Barnsley and Aristid Lindenmayer found that the inclusion of this final symmetry brought with it an effective description of the natural world. The *Koch curve* contained the quality of roughness seen in cracks and coastlines [1], the Barnsley fern closely resembled real ferns [2] and Lindenmayer's L-systems generated visually plausible plants and trees [3, 4]. The work by Benoit Mandelbrot in particular emphasised that the inclusion of scale symmetry introduces roughness as a fundamental component of geometry rather than a deficiency in an object, as it had been perceived [5].

This late adoption of scale symmetry is still challenging our traditional mental model of the world. For example, the asteroid belt in our solar system is said to be composed of one dwarf planet, many asteroids, millions of meteoroids of less than 1 m, and lots of dust. However, the whole belt is suggestive of a scale-symmetric system, it follows a *power law* where there are approximately 750,000 $n^{-1.9}$ bodies of more than n km in diameter. If the spacing between asteroids scales similarly, then it has *statistical scale symmetry*, and from a scale-symmetric viewpoint, we can consider it a single structure.

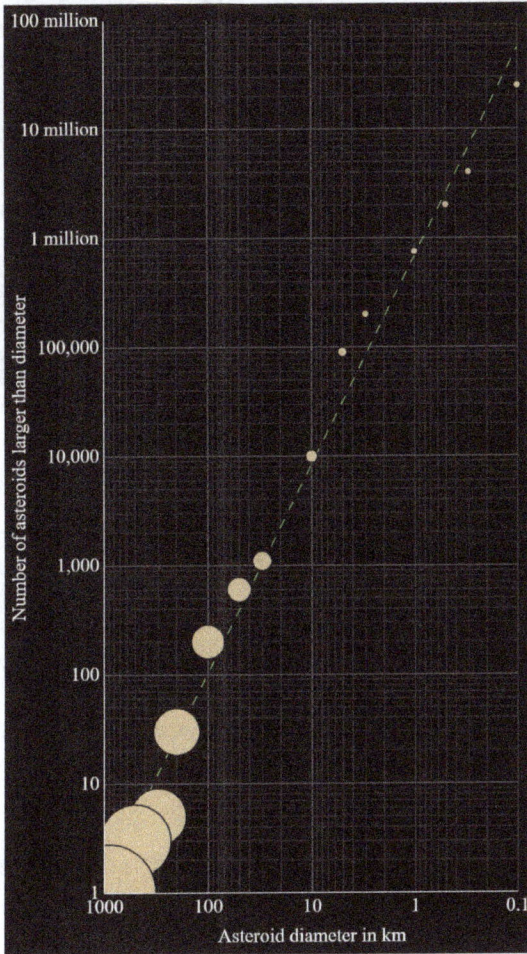

Figure 9.   The distribution of asteroid sizes is approximately scale-symmetric — the relationship between diameter and the number of larger asteroids is independent of scale. This power-law relationship follows an approximately straight line on a log–log plot of the two parameters [6].

There is a similar case for forests. In the 2008 paper "A general quantitative theory of forest structure and dynamics" [7], West *et al.* found that the power law that described the size of branches on a tree also described the size distribution of trees of all sizes in a forest. This means that, from the perspective of scale-symmetric geometry, a forest can be thought of as one large tree-like structure rather than a complicated distribution of trees and saplings of different sizes. This idea of a forest as a single structure will be seen again in Chapters 3 and 8.

The most well-known example though is the coastline of Great Britain. In 1961, Richardson [8] showed that the length of the coast appears to grow as you use smaller yardsticks to measure it, in fact the length approximately follows a power law with respect to the stick length. This coastline follows a statistical equivalent of a scale-symmetric curve, such as the Koch curve. So, rather than seeing the coastline as being an imperfect version of a complicated polygon, from the perspective of scale-symmetric geometry, it is a single curve defined by a power law.

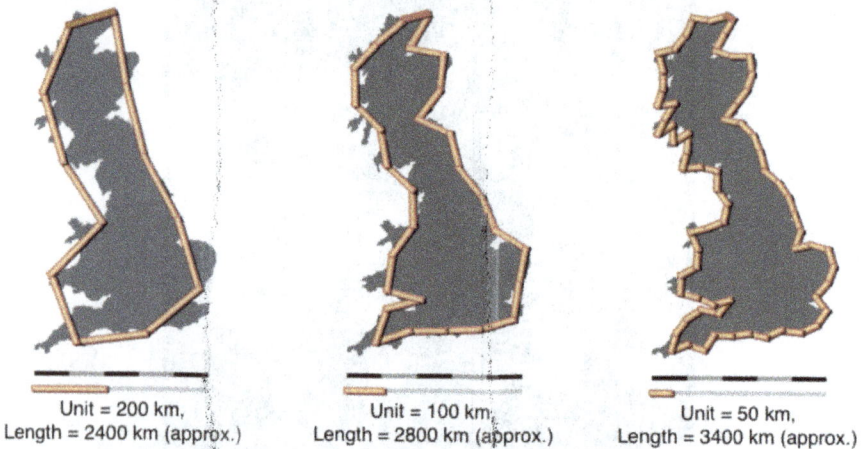

Unit = 200 km,
Length = 2400 km (approx.)

Unit = 100 km,
Length = 2800 km (approx.)

Unit = 50 km,
Length = 3400 km (approx.)

Figure 10.   The coastline of Great Britain does not converge when measured with smaller yardsticks. Instead, it follows a power law, which means that the coastline is a statistically scale-symmetric, fractal curve.

This *power law* defines the total length L of the coastline as a function of the length l of the measuring stick: $L(l) \propto l^{1-d}$, where the symbol $\propto$ means "is proportional to." The parameter d is known as the *fractal dimension* of the coastline, in this case, it describes its roughness. Smooth coasts have $d = 1$ and the value tends towards 2 for increasingly rough and convoluted coastlines. The value for Great Britain is approximately $d = 1.2$.

In general, the *fractal dimension* of a shape describes the rate of growth of occupied space with respect to resolution (m). Lines grow with m^1, planes grow with m^2 and general shapes grow with m^d where d is the *fractal dimension*. There are several ways to calculate d, the method above is called the *yardstick method*, but a more general technique counts the cells of an $m \times m$ grid that overlap the shape. Its output is called the *box-counting dimension* of the shape.

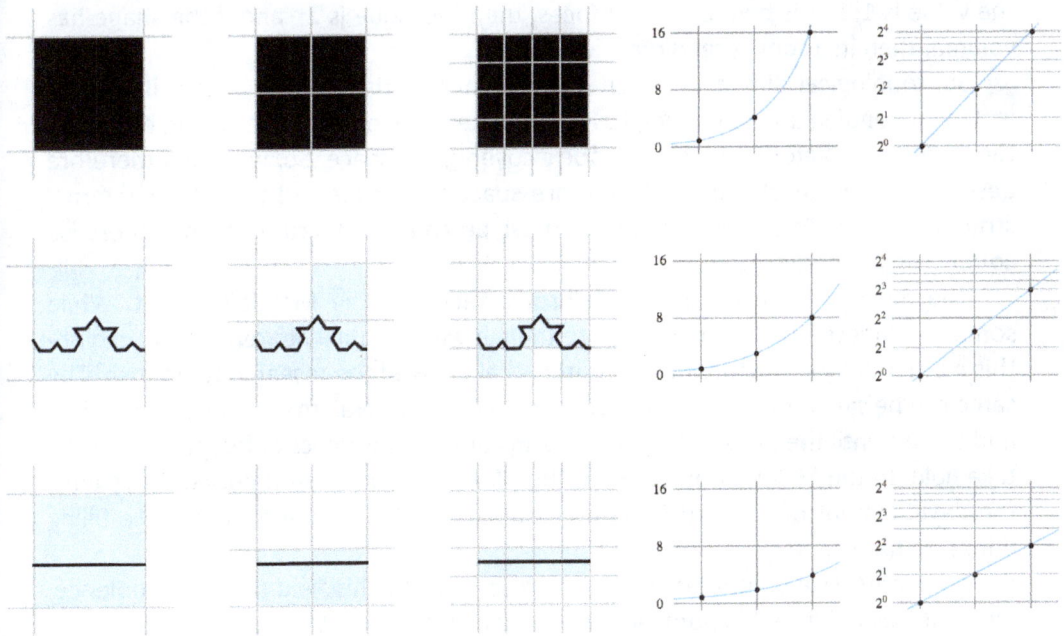

Figure 11. Box-counting dimension on a square region (top), a Koch curve (middle) and a line (bottom). Plotting the number of pixels covered, with respect to each doubling of the resolution *m*, a curve (second from right). Using a log range (right), the box-counting dimension is the average gradient of this line.

We will discuss this and other types of fractal dimension in Chapter 9, but for most of the structures in this book, the different types are equivalent and we can characterise them by a single *fractal dimension*. These *fractal dimensions* were described by Benoit Mandelbrot in 1967 [9], and their use was popularised in the 1980s with his research on *fractals* [5].

Fractals are a special type of scale-symmetric shape. The term *fractal* has a loose meaning in common speech, but for this book, we will adopt the standard mathematical definition which is:

Fractal: A subset of Euclidean space with fractal dimension greater than its topological dimension.

Here, a "subset of Euclidean space" just means a shape in some *n*-dimensional flat space. *Topological dimension* (or *Lebesgue covering dimension*) is an integer value, with a quite nuanced definition [10], but it is usually easy to pick its value. If the shape is made up of points, then it is 0; if it is made up of lines or curves, then the value is 1; if it is built out of surfaces, then the value is 2; and if the shape has volume, then its topological dimension is 3.

So, topological dimension is an internal characteristic of the shape, it is unaffected by bending and stretching it. Fractal dimension, on the other hand, describes the growth characteristics of the shape's coverage of space. So, we could therefore describe fractals as shapes that fill more space than expected from their internal structure alone. To achieve this, they must be contorted, crumpled or otherwise complex in appearance.

The most well-known of Mandelbrot's fractals is the Mandelbrot set. While some fractal research took place before then, it was the ability to render and explore this mathematical structure on computers that allowed the research to flourish. The same can be said for the study of scale symmetry in general, the research in the field had to wait until the personal computer and computer graphics before it could really take hold. In the 1980s alone, it led to the *Mandelbrot set*, the theory of L-systems, the development of *Iterated Function Systems*, the Barnsley fern and space-filling tilings, to name a few.

The impressive computer renderings also brought this field to wider audience, where it was published in popular titles, such as *Fractals Everywhere* [2], *The Fractal Geometry of Nature* [5] and *The Beauty of Fractals* [11].

> *The advent of the computer, not as a computer but as a drawing machine, was for me a major event in my life.*
>
> — Benoit Mandelbrot

We are experiencing a second wave of interest today. The development of General-Purpose Graphical Processing Units together with web technologies such as webGL has allowed 3D scale-symmetric structures to be visualised in real time. Even more importantly, social media, forums and other online communities have created an unprecedented ability to collaborate across disciplines. This suits the topic of scale symmetry well, it is a topic that stretches beyond academics and is used by programmers, artists, hobbyists, architects and game designers among others. In recent years the online communities have combined rendering software from interested programmers with the formulae from mathematicians and the eye for composition and colour from artists to produce impressive 3D structures and scenes. Some of these are shown at the end of Chapter 10. Several programs and fractal structures have even featured in major film works in recent years.

Due to its relative youth, this is a very open field of study. There are many unknown facets, unproven ideas and mathematical structures that have not been seen. With the aid of online information, software and communities, anyone can explore the space and see something new. All they need is to have an idea of where the interesting sorts of structures lurk and how they are constructed.

This is where this book comes in, the following six chapters are an exploration into the depths of scale-symmetric geometry. Starting with fractals, then non-fractals, sphere inversions, Mandelbrot sets, cellular automata and finally scale-symmetric dynamics. In each case seeking unusual and novel examples, and characterising and comparing their properties.

But an exploration is more than producing specimens, it should survey and map the landscape. I address this in Chapters 8 and 9 where I discuss a classification system for these geometries and quantification methods, including generalising the notion of fractal dimension beyond the positive real numbers. Together these allow the broad set of geometries to be organised and ordered, and these systems are used to provide a basic survey of the field.

So, let's get going! Casual readers can skip past the blue information panels without issue. On the other hand, determined readers can find the software and algorithms used in this book cited in Chapter 10, where examples can be built and modified. As you read, make note of the use of *italics*, these are specially defined terms and labels that you can refer to in the glossary at the back of the book.

Chapter 2

Fractal Structures

Fractals are, as they say, everywhere. In clouds, mountains, cracks and coastlines. But also scattered through internet forums, phone apps, maths magazines, student posters and as a background image to any site that aims to look advanced or new age. They are popular and popularised and so what people think of as fractal has become quite broad. However, mathematically, fractals are a very specific class of shapes and represent only one special type of scale-symmetric structure.

So, in this chapter, we look at what fractals really are and how to construct them using one of the simplest examples — the Koch curve. From here, we begin exploring variations and then generalisations of this important curve into three-dimensional (3D) surfaces. We continue by considering more complex structures than just curves and surfaces, such as trees and sponges and then finish by looking at a way to classify these different types of fractal according to their structure.

Let's begin by looking more closely at the definition of the term *fractal,* as given in the introduction:

Fractal: *A subset of Euclidean space with fractal dimension greater than its topological dimension.*

This *fractal dimension* has several types, and we mentioned the *yardstick* and *box-counting* dimensions in the introduction. The most general of these is the *Hausdorff dimension,* and this is the usual choice for the above definition.

Its definition is difficult, but fortunately for simple scale-symmetric shapes, these different dimension values coincide, so we needn't worry about this trickier case. The most straightforward type is the *similarity dimension*. For basic fractals, it is easy to calculate, so let us try it out using the *Koch curve* as an example. Its simplest construction is to repeatedly substitute a single oriented line segment with two smaller segments in a V shape, with their orientation flipped.

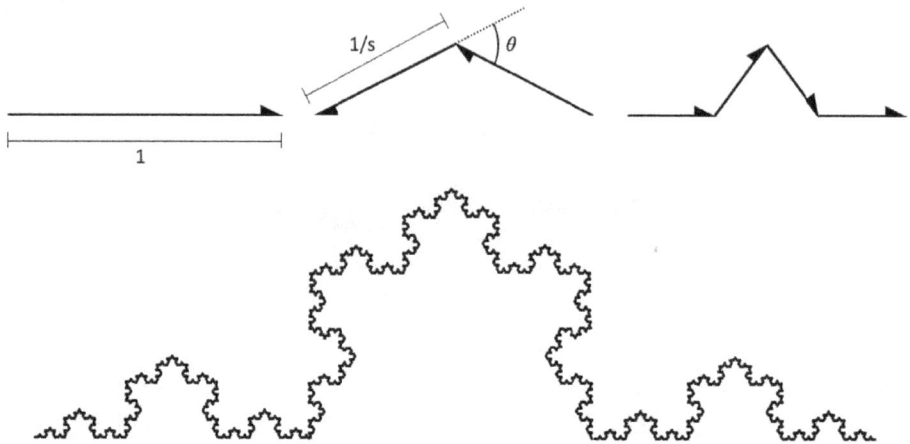

Figure 1.   Koch curve construction, using one-sided arrows to represent the oriented line segments. Top: first three iterations $i = 0,1,2$. Bottom: limit set as $i \rightarrow \infty$.

This process is an example of a *substitution rule*. These rules tell you how to replace a shape primitive with an arrangement of n smaller copies. The process is repeated every iteration i and the resulting shape as $i \rightarrow \infty$ is called the limit set. This limit set is the fractal that is generated from the substitution rule.

In this case, the shape primitive is the oriented line segment, and the $n = 2$ smaller copies are each $1/s$ of the parent's length. This is sufficient information to measure the *similarity dimension* of the resulting fractal, it is $D = \frac{\log n}{\log s}$. This makes the standard Koch curve approximately 1.26-dimensional. Since this is greater than the topological dimension of a curve (which is 1), we can see from our definition that the Koch curve is a fractal.

We can parameterise this curve by its bend angle θ, which affects the scaling s such that its fractal dimension can take any value from 1, when the bend angle is 0, to 2 when the bend angle is 90°. This makes it a very neat example of a fractal curve.

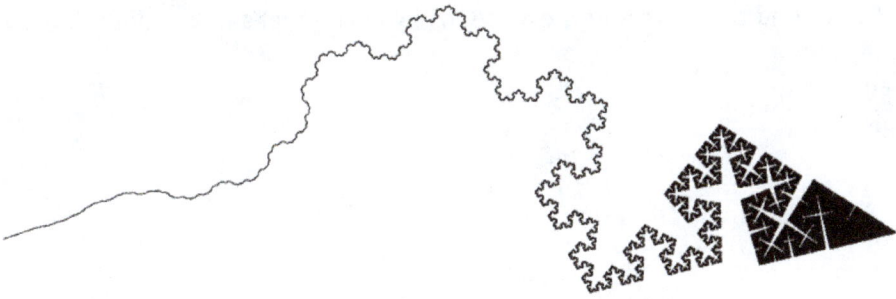

Figure 2.   Von Koch curve with bend angle changing from 0° to 90°.

Even with this simple construction, there is some exploration that can be done. A quick calculation shows that there are six distinct permutations of the line replacement orientations if we include reflections. The first three are well known: the *Levy C curve*, the *Koch curve* and the *dragon curve* families:

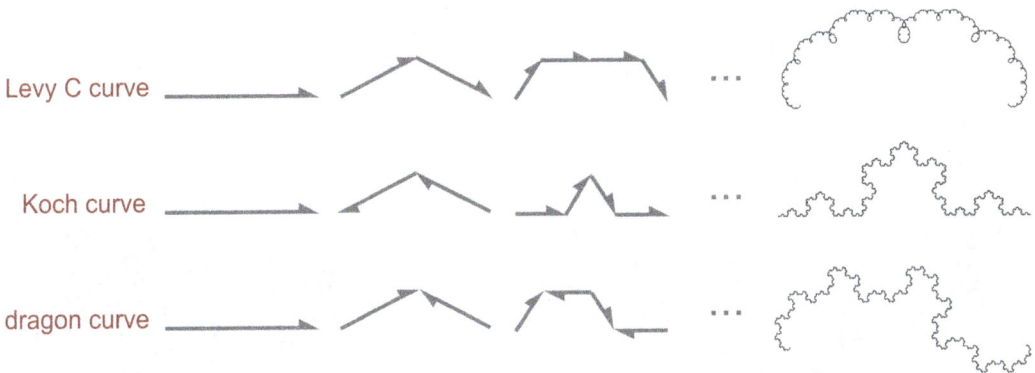

Figure 3.   Iterations $i = 0,1,2$ and the limit set $i \to \infty$ for the well-known Levy C, Koch and dragon curve families. Shown for 60° bend angle, though the typical Levy C and dragon curves use 90°.

The second three are barely ever seen however, so let's take a closer look.

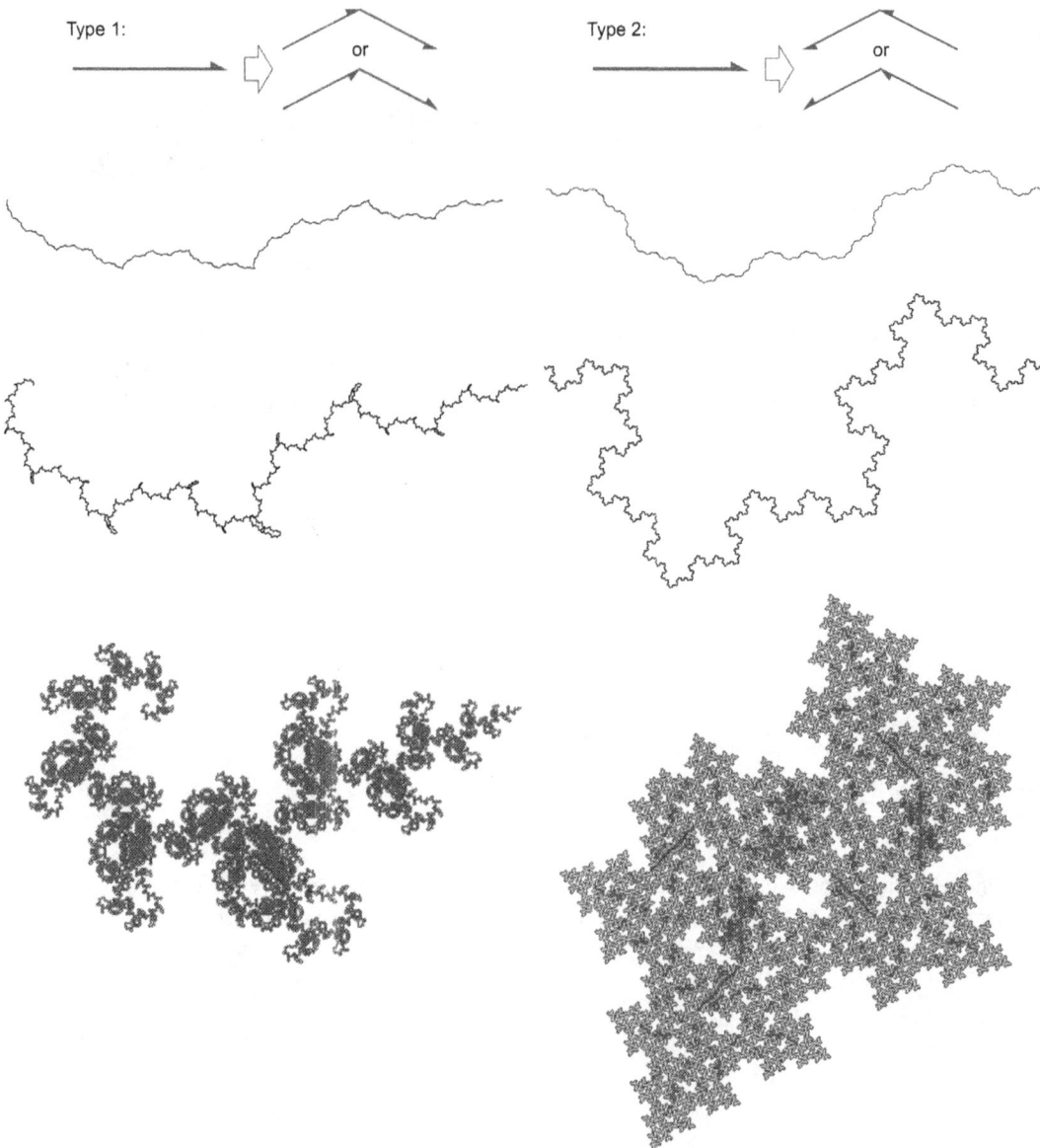

Figure 4.   Curves for types 1 and 2 substitution rules, shown with increasing bend angle from top to bottom.

Type 3:

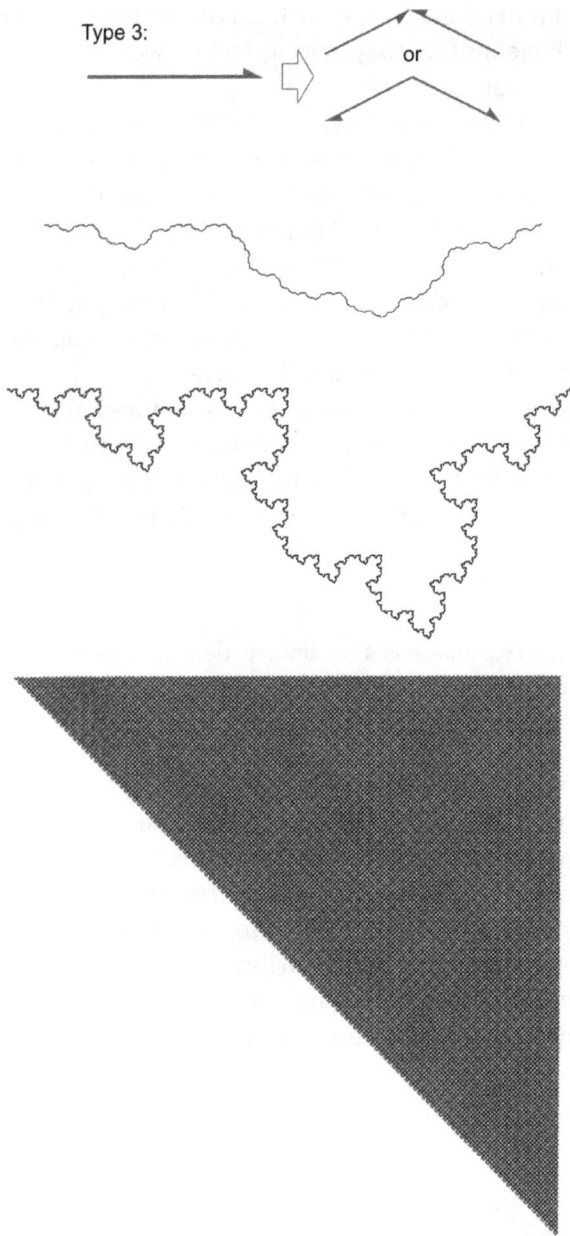

Figure 5.   Curves for type 3 substitution rule with increasing bend angle. Bottom: 90° bend angle fills the plane.

It is interesting to see that this last example (type 3) can also transition between a line and a solid area without self-intersection.

At the maximum bend angle, the Koch curve and type 3 curve are both *space-filling curves*, which means that they densely fill the two-dimensional (2D) space and so have fractal dimension 2.

Most of these fractal curves are *embedded in* 2D space, which means they can be defined as just a set of 2D points. However, some of the fractals self-intersect, like the *Levy C curve*. These are said to be *immersed in* 2D space because the overlapping points need to be included. This can be achieved by defining them as a *multiset* of 2D points, where the overlapping locations are counted multiple times. The act of embedding these fractals reduces them to a set, and the multiple coincident points are lost. In doing so, the dimension of the fractal is often reduced below that calculated from the *similarity dimension* formula above.

These topologically one-dimensional (1D) structures of course exist in three dimensions as well, but since their construction is quite similar, let us instead fast-forward to a more difficult construction: fractal surfaces. Is there something of equivalent simplicity to the *Koch curve* in the form of a surface in 3D? More specifically:

> Is there a fractal surface that has a continuous parametrisation between a flat plane and a solid volume, while never intersecting?

It seems that would make an excellent archetype against which different rough surfaces could be compared. Unfortunately, it is still an open question whether such an ideal surface exists, but we can get quite close with the following constructions.

The *pyramidal surface* fractal uses a *substitution rule* just like the Koch curve. Rather than replacing a line segment with two smaller line segments in a V shape, it replaces one triangle with six smaller triangles of which four are in a pyramid shape. The steepness of this pyramid is equivalent to the bend angle in the Koch curve. The difficulty comes in calculating the correct triangle shape.

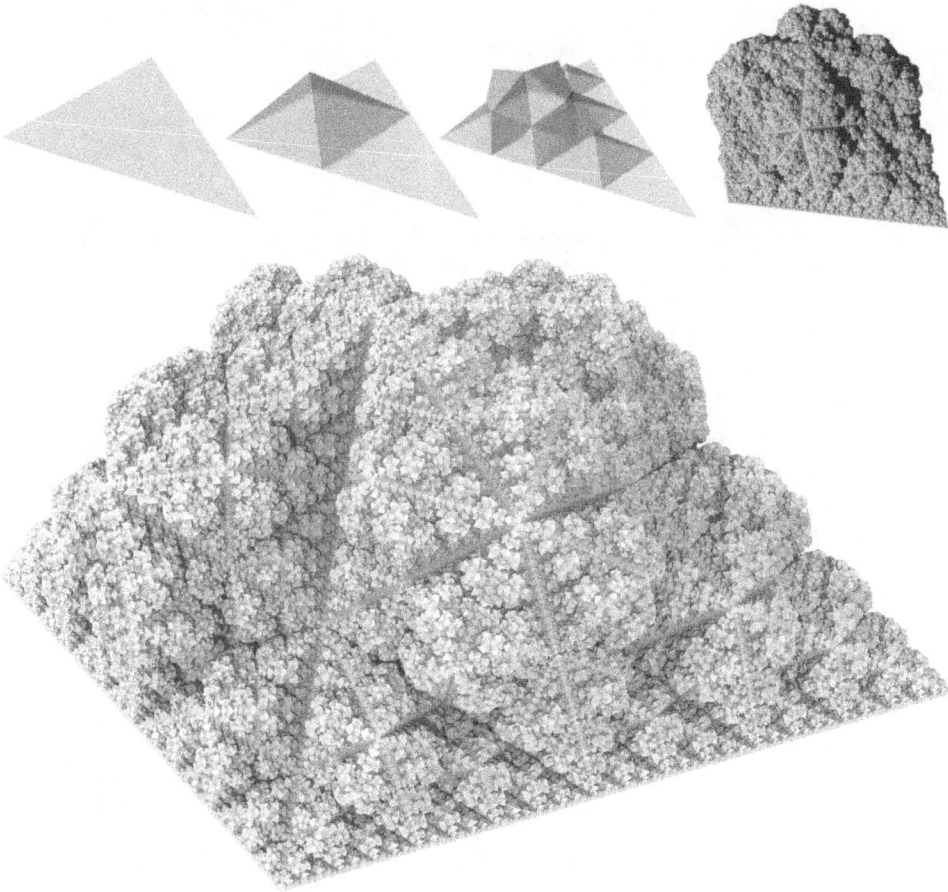

Figure 6.   Top: *pyramidal surface* fractal, iterations 0,1,2 and ∞ (for just the pyramid section), with fractal dimension approximately 2.3. Bottom: close up.

Construction

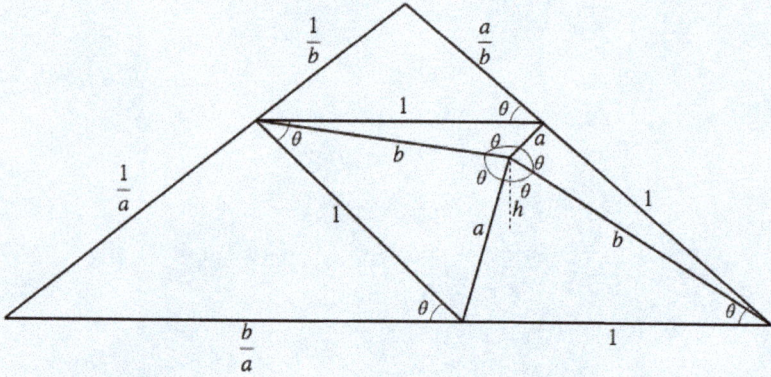

For pyramid base edge length 1 and acute angle θ of the base diamond, the pyramid height is:

$$h = \sqrt{\frac{1+\sqrt{1+\tan^2(\theta)-\sin^2(\theta)}}{2\sec^2(\theta)-2}}$$

which is sufficient to label the other triangle lengths as:

$$a = \sqrt{\sin^2(\theta/2)+h^2}$$

$$b = \sqrt{\cos^2(\theta/2)+h^2}$$

The similarity dimension D is now the solution to the equation:

$$a^D + b^D + 4a^D b^D = (a+b)^D$$

When the base shape is a 45° right-angled triangle, the surface is a flat plane, and as the triangle becomes more acute, the pyramid pitch increases and the surface becomes rougher. However, the surface self-intersects before becoming a 3D space-filling surface. It intersects when the surface is approximately 2.3D and very close to the point where the base triangle becomes isosceles. The image in Figure 6 is at this contact point. The fractal overlaps itself at higher bend angles and so its fractal dimension can reach and even exceed three. It is only when the overlapping points are removed (when we embed it) that the fractal dimension is bounded to 3.

Figure 7. *Pyramidal surface* fractal, with pyramid height *h* from –0.2 to 0.6.

As can be seen above, this surface contains several distinct stages, for downward pyramid directions it is characterised by a hierarchy of basins, for 45° right-angled triangles it covers a flat plane, and for upwards pyramids it is characterised by a hierarchy of hills. At the point that the surface self-contacts, it represents the surface of a sponge-like shape, as its structure contains connecting loops.

The downside of this variety is that, unlike the Koch curve, this surface looks different when flipped over and seen from the other side. In order to replicate the "flip symmetry" of the Koch curve, we could flip the direction of some of the child triangles in the construction, however there isn't an obvious choice of which to flip that would be balanced and symmetrical.

Another property of the Koch curve is that decreasing the bend angle acts like pulling a crumpled string back into a taut and straight line. The equivalent for a surface would be that any rough surface could be pulled back to a flat plane. If, like the string, we assume that the surface to be incompressible then it ought to act like crumpled paper. Mathematically, this is called a *developable surface*; it can unfold to a plane without stretch or compression. Could we generate such a surface?

We are now asking for a lot:

Is there a simple, scale-symmetric and flip-symmetric developable surface that can transition from 2D to 3D through its single bend parameter without self-intersecting?

This would indeed be a wonderful object, it would in a sense be the ideal way of crumpling paper and could be a useful model for other surfaces that crumple such as mountains perhaps. Unfortunately, this is also an open question.

However, if we relax the final constraint and allow a small amount of self-intersection, then the following method is effective. Let us call it the *crumpled surface fractal*.

We start by noting that there is a different way to construct the Koch curve. Starting with a horizontal line: tip it and reflect between two horizontal mirrors to generate a zig-zag, now rotate the tip direction 180° around the vertical axis, double the mirror spacing and repeat.

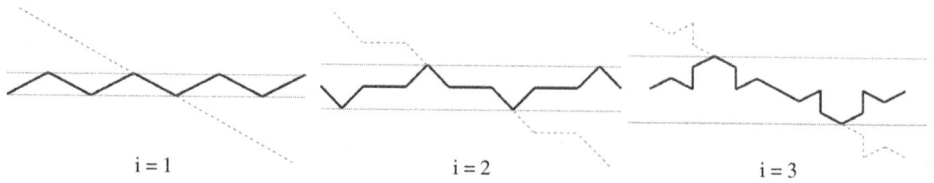

Figure 8.   The first three iterations of an alternative method to generate the Koch curve. Starting with a horizontal line: tip clockwise (dashed line), reflect between red mirrors to give black line, then rotate 180° around the vertical axis.

After applying this iteration many times, we see the Koch curve start to form. In fact, it forms two Koch curves, one to the left and one to the right of the centre of rotation.

We can apply this method to a horizontal plane rather than a line, and it produces extruded Koch curves instead. To generate the *crumpled surface fractal*, we just need to rotate the tip angle 90° rather than 180° and scale the mirror separation by the square root of two instead of by two. Therefore, two iterations in the planar case take it to the same state as one rotation in the linear case.

The value of using mirrors to generate the surface is now evident because the mirror transform always maintains the developability of the surface. It will always unfold into a flat plane.

Figure 9.   Top: the first five iterations of the mirror-and-rotate method. Bottom: the limit sets for increasing bend angles.

The result is an interesting crumpling method that appears to crumple the paper into a pattern of square spirals. The larger tip angles produce a darker surface because light is scattered more by the stronger undulations and more areas become shadowed.

As well as the tip angle, we can also parametrise the surface by the rotation angle around the vertical axis, which is the yaw angle y. The mirror separation scale is then $2^{y/180}$.

Figure 10.   Varying the rotation angle. From left to right: 180°, 90°, 120° and 60°.

For 60° yaw angle, the square spirals become hexagonal spirals and the surface starts to look a little less artificial. In that vein, a yaw angle of 180° divided by the golden ratio is a nice choice to minimise the alignment of tip directions as we change the yaw. It acts to avoid the repeated rotations generating a regular pattern such as the square or hexagonal spiralling in Figure 10. If we view a section of this surface away from the origin, then the crumpling effect is quite random and the surface looks remarkably similar to crumpled paper.

Figure 11.   Crumpled surface with yaw angle 180°/ϕ (the golden ratio) shows a more natural crumple effect. The value $\phi \approx 1.618$ avoids the multiple rotations forming a repeating pattern.

A nice property of a fractal developable surface is that it can be textured without distortion.

Figure 12.   Texturing the fractal surface with an image of music on old paper. The musical notes are not stretched or distorted, only crumpled.

It also makes a reasonable approximation to mountainous landscapes that are created through moving and colliding plates. Unlike volcanoes, the mountains in this scenario are not a series of protrusions, and not like the hills-on-hills structure of the previous *pyramidal surface* fractal, nor are the valleys a hierarchy of basins. Instead, the landscape is defined by ridges that sometimes branch and winding valleys that also sometimes branch.

We can make use of these properties to make a reasonable attempt at a landscape, using a texture of trees on rock, with a road, and rendering it for increasing tip angles. The results are quite effective, with small tip angles appearing like hilly terrain and large tip angles giving a much more craggy and inaccessible jungle sort of impression. The road texture does not translate too well of course, as real roads would cut through the hills to remain flat.

Figure 13.   A forested terrain image applied to the *crumpled surface* fractal for increasing bend angles. The impression is of a mountainous terrain that gets increasingly craggy.

The similarity dimension of the surface is a simple function of just the tip angle ϕ:

$$D = 2\log_2 \frac{2}{\cos\phi}$$

While this crumpled surface fractal has a simple dimension formula, it is not in fact built of a simple pattern of triangles, and this pattern varies with the tip angle. So, it cannot be generated using a simple substitution rule as with the previous surface, and the mirroring operation that it uses can be quite computationally costly. It also contains some self-intersection, which is minimal for small tip angle but starts to become visible for the larger tip angles presented here. Nevertheless, it makes for an attractive fractal surface in both the stochastic golden ratio form and the Koch-like, right-angled form. Also, the surface is in some sense doubly fractal: not only is the surface a fractal, but every path on the surface is also a fractal curve, so it would take forever to travel between any two points. This is unlike the pyramid surface, where any two points are connected by a finite length path.

★

The value of these variable-dimension curves and surfaces is that we can produce just the roughness that we need. For instance, we can make a 1.5D Koch curve, which is half way between being a 1D line and a 2D plane.

If we don't require a specific fractal dimension, then 1.5D is a nice choice to demonstrate a fractal curve, as it is farthest from the two Euclidean shapes at 1D and 2D, so it is in an approximate sense the "most fractal" curve. While a 1.99D Koch curve has a higher fractal dimension, it doesn't visually appear rougher or much like a fractal as it is very nearly a solid square, just with some small cracks in it.

In general, fractals are farthest from Euclidean when their fractal dimension is halfway between their topological and embedding dimension (the dimension of the space they're embedded in). We can call these *semi-dimensional shapes,* they make excellent example fractals, and fortunately their construction is often simple.

When the difference between embedding and topological dimension is even, then the *semi-dimensional shape* has an integer dimension. This does not mean that it is Euclidean, for example, the following 2D curve embedded in three dimensions. It is generated just like the Koch and Levy C curves, but with the child line segments rotated 90° around their axes in each substitution:

Figure 14.   Top: first three iterations of 2D Koch curve in 3D space. Bottom: resulting limit set.

When the difference between embedding and topological dimension is odd, then the *semi-dimensional* shape will have a half-integer number of dimensions. We can generate these fractals using simple right-angled geometry by dividing the base shape lengths by four and building each iteration out of a number of these child pieces. For a 1.5D curve in 2D space, we get an odd looking mushroom cloud sort of shape:

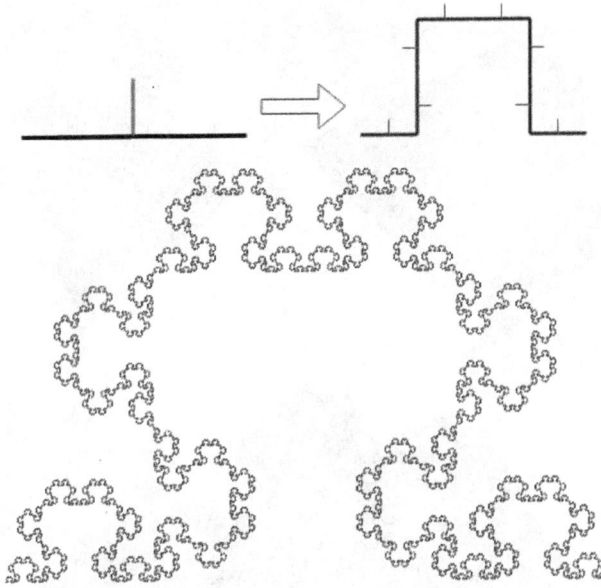

Figure 15. Top: first two iterations of substitution of one line segment with eight segments of one quarter the length. Bottom: the resulting *semi-dimensional* curve.

and for a 2.5D surface in 3D space, we get an equivalent "mushroom cloud" surface:

Figure 16.   Top: first two iterations replacing each square plane with 32 planes of one quarter the side length. Dark planes are those with orientation flipped. Bottom: resulting *semi-dimensional* surface.

Neither of these examples are unique, but they are both the cases with fewest concavities while avoiding self-intersection. What's more, this semi-dimensional surface is the highest dimension non-intersecting surface that we have found.

These examples are geometrically complex, but they remain topologically simple, having the topology of simple curves and surfaces. We can say that these fractals are topologically equivalent to a line or plane; they are just a contorted line or plane. But there are many other structures possible, a very common example are trees. Let us look at those next.

Fractal trees can be generated with a replacement method similar to the curves described previously. First, we take a star of n radial lines and we iteratively scale down and attach n identical shapes to its outer points, such that the connecting line is straight.

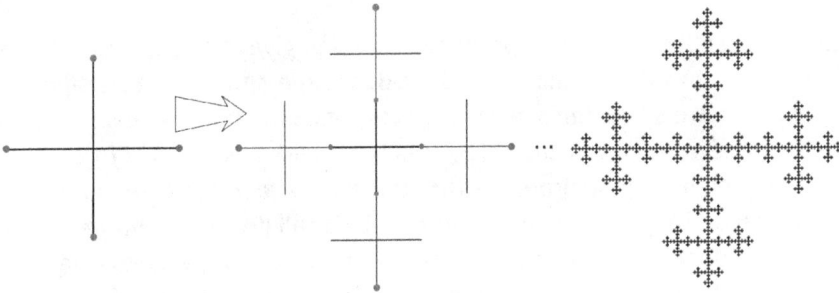

Figure 17.   Fractal tree iterations 0, 1 and ∞ for $n = 4$: four copies are attached at the coloured outer points and four new outer points chosen.

When n is odd, the outer point is ambiguous, so we apply a consistent rotation direction to generate the new shape and its outer points.

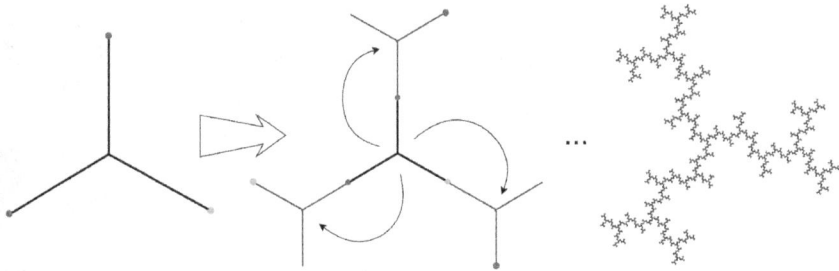

Figure 18.   Fractal tree iterations 0, 1 and ∞ for $n = 3$. The central shape is rotated around each coloured pivot and generates each new pivot point.

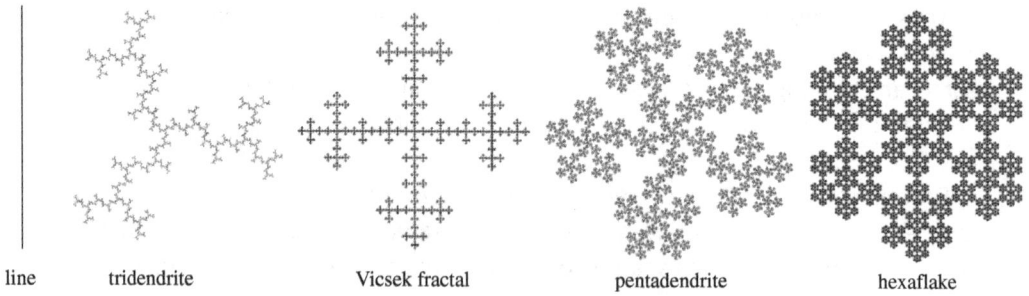

| line | tridendrite | Vicsek fractal | pentadendrite | hexaflake |

Figure 19.   The resulting dendrite family, $n = 2$ to $n = 6$ from left to right.

The $n = 2$ case is too sparse to produce a tree and gives a straight line. Conversely, the $n = 6$ case is too dense and instead produces the sponge-like *Hexaflake*, these fractal networks are sometimes also called carpets, like the square-shaped *Menger carpet*. The in-between $n = 4$ case produces the well-known *Vicsek fractal* tree. For odd numbers, the $n = 5$ case produces the *Pentadendrite* fractal tree, but there is not a known name or image for the $n = 3$ case, so let's call it the *Tridendrite* fractal tree.

While the *Tridendrite* is highly geometric in its form, it is perhaps a better archetype than the others for many trees in nature, since three-way branch points are the general case, as they don't require coordinated branching. As such, phenomena like lightning, cracks, trees, and rivers usually have these Y-shaped forks.

These examples are starting to show a richer space of different fractal structures. Not just trees but sponge-like structures, such as the *hexaflake*. Here are two other structures, both with variable dimension:

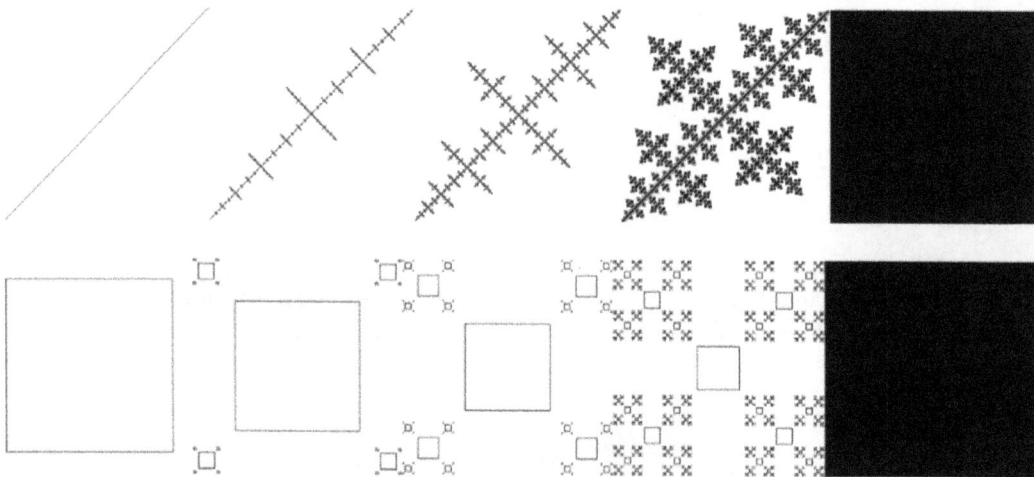

Figure 20.   Two variable-dimension fractals from 1D (left) to 2D (right), with non-trivial topology.

There are clearly a lot more types of structure than just fractal curves and surfaces, so what others are out there, and how do we find them? Let's try and answer these questions.

An effective search would be to employ a general-purpose fractal construction method that is simple enough to allow exhaustive cataloguing of its fractals. A neat method for this is to apply cell replacement on a 3×3 grid. The grid cells are labelled as *c*: centre, *e*: edge, *v*: vertex, and for each label, you either replace the respective cells with a 3×3 child grid, or they remain empty. For example, the structure generated from a "*c*" construction replaces only the centre cell with the smaller 3×3 grid, the other cells are empty. This is repeated and at its limit, there is only a single point in the centre that is not empty.

This recursive definition has a binary choice for each label, so there are eight resulting fractals:

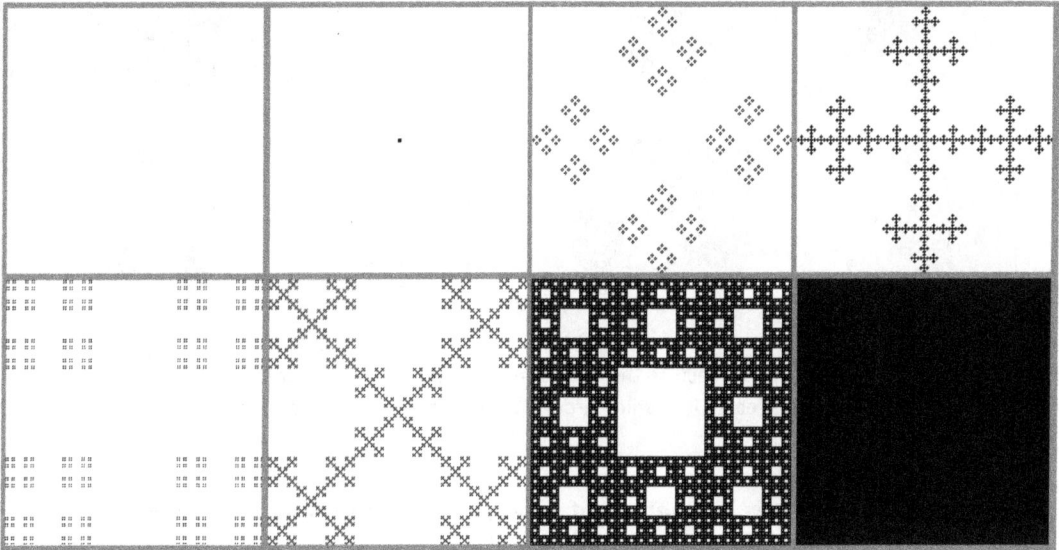

Figure 21. 3×3 grid fractals recurse on the centre (*c*), edge (*e*) and vertex (*v*) cells, we label the recursing cells of each, in reading order: –, *c*, *e*, *ce*, *v*, *cv*, *ev*, *cev*.

The resulting shapes are familiar, they include *2D Cantor Dust*, the *Vicsek* fractal and the *Menger carpet*. But beyond looking at their exact geometry, we can see several structures: two tree shapes, a sponge-like structure and two clusterings of disparate points. So, this method repeats the previously seen structures and introduces the point clusterings as an additional basic type.

The method can be applied in higher dimensions as well. In 3D, it is applied to a 3×3×3 grid, with four labels *m*: middle, *c*: face centre, *e*: edge and *v*: vertex, which produces 16 structures:

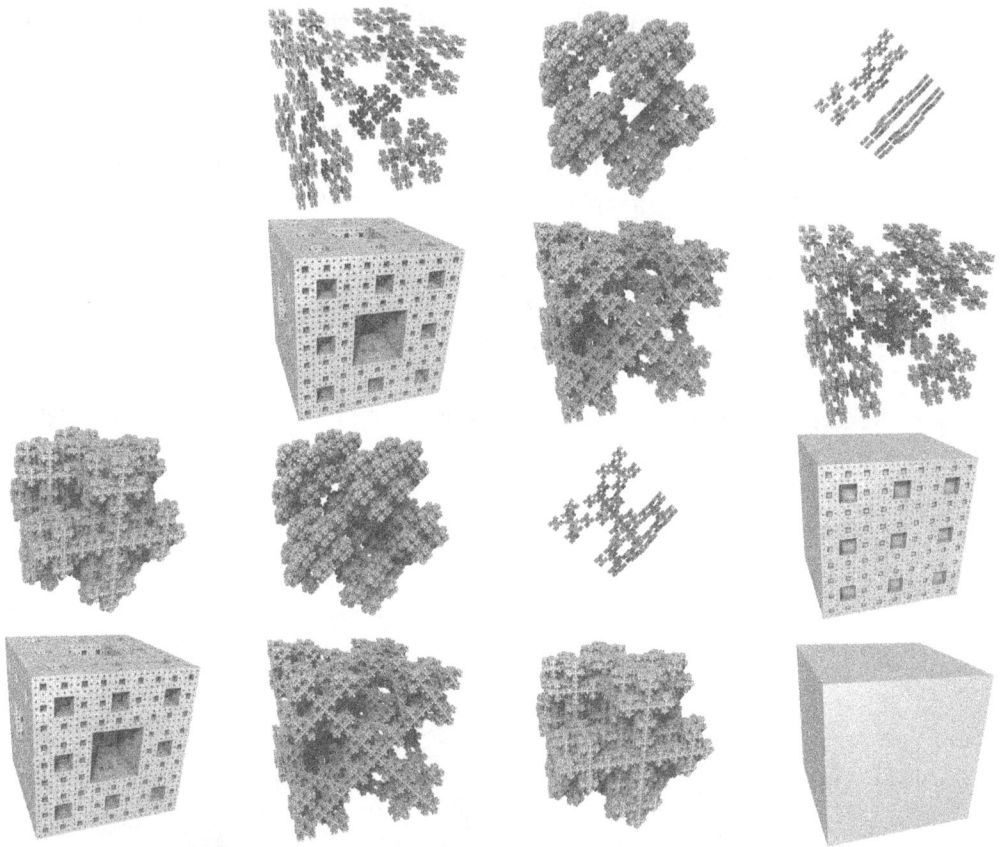

Figure 22.   3×3×3 grid fractals recurse the middle (*m*), face centre (*c*), edge (*e*) and vertex (*v*) cells. We label the recursing cells of each, in reading order: –, *v*, *e*, *c*, *m*, *ev*, *cv*, *mv*, *ce*, *me*, *mc*, *cev*, *mev*, *mcv*, *mce*, *mcev*.

In this case, we see a larger set of structure types. In addition to the 3D cluster and 3D tree types, we see a 3D sponge-like fractal — the well-known *Menger Sponge* — and what I would call a foam-like fractal. This fractal, labelled *cev*, is dividing a solid cube into a 3×3×3 grid and repeatedly removing the middle cell, which means that it is built of many disconnected "air" pockets. This sort of fractal is approximated in the real world in foam and bubbles on the surface of a liquid. The final non-trivial structure (*mce*) is similar to a tree but appears as a set of recursively branching planes. From an exploring point of view, this last type is interesting because it is rarely seen in fractal examples.

Without a formal definition of "basic structure types," it isn't certain that this cell-based method captures all the basic types. Nevertheless, it has a very general pattern to it, where the connectivity of the structure is determined by which fundamental pieces of the structure you remove at each iteration: the corners, the edges, the faces or the volumes — a list that continues in higher dimensions. As such, it is a very useful tool in finding the most important fractal structures.

Later on, in Chapter 8, this tool and the structure types seen here will form the basis for a full classification system. This is useful for spotting the unexplored regions, where new structures are waiting to be found!

But before then, we will begin our exploration of *general* scale symmetry. We have so far only touched the surface of scale-symmetric geometry, most of which is not fractal by our definition. So, in the next chapter, we will specifically look at non-fractal geometries, and we'll find that there is far more to explore than just fractals.

Chapter 3

Non-Fractal Structures

There are many scale-symmetric structures that are not fractals. A common example are recursive shapes built from progressively smaller copies of a base shape. Each recursion the child shapes add to (rather than replace) the parent shape, a so-called *multiplication rule*, rather than *substitution rule.* In these shapes, it is only a component of the structure that is fractal. For recursive tree shapes, it is the end points (leaves) of the tree that are the fractal component.

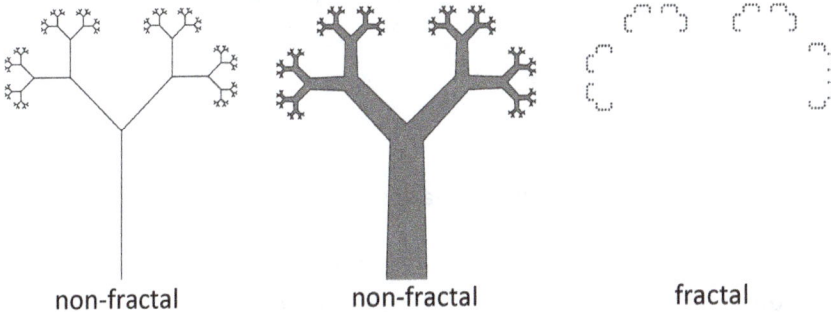

| non-fractal | non-fractal | fractal |

Figure 1.   These binary trees have equal fractal and topological dimensions (left: 1, middle: 2), so are non-fractal. However, their end points are fractal, with fractal dimension (1) greater than the topological dimension (0).

Where the branches have a thickness, it can also be the outline of the trees that are the fractal component, these thick trees are also non-fractal as their topological and fractal dimensions are both 2 (in 2D). There are very few examples of such trees with a fractal outline, so let us build one and, where possible, let us try to add some realistic rules in its construction.

The main decision in making such a tree is to decide how it branches. In particular, what are the thicknesses and angles of the two child branches relative to their parent? Leonardo da Vinci had a nice rule for this. In his notebook, he wrote:

All the branches of a tree at every stage of its height when put together are equal in thickness to the trunk.

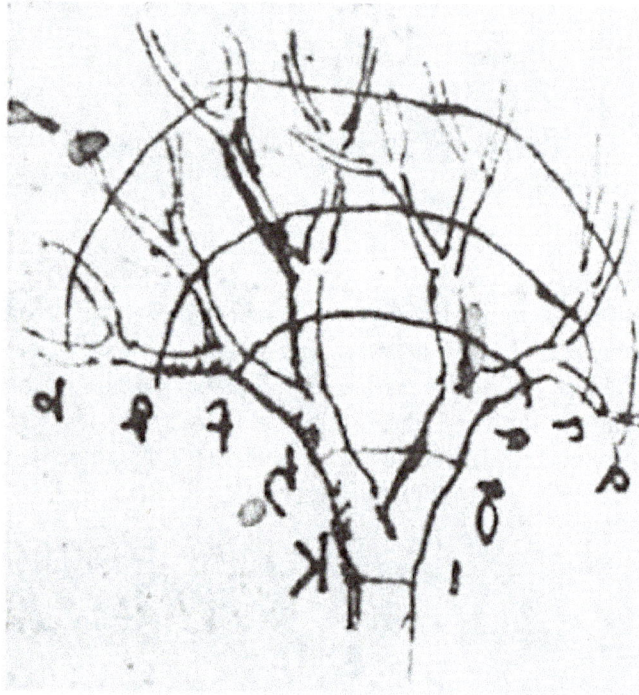

Figure 2.   Depicting the tree-branching rule by Leonardo da Vinci.

This means that for any one branching point, the square of the parent branch radius equals the sum of the square of the children radii. If we wish the branches to exit straight out from the junction, then the consequence is that the junction must be a right-angled triangle in 2D.

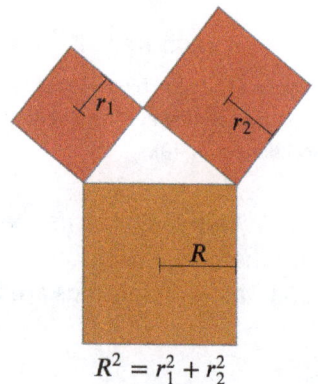

$$R^2 = r_1^2 + r_2^2$$

Such trees in 2D are called *Pythagoras trees*.

Figure 3. A Pythagoras tree for a particular right-angled triangle. The squares are coloured by the iteration they are added in.

However, it seems unlikely that Leonardo's rule for three-dimensional trees would apply equally in 2D, and even in three dimensions, his rule is only approximate, in *Leonardo's Rule, Self-Similarity, and Wind-Induced Stresses in Trees* [12] by Christophe Eloy, it states that:

> *For real trees, the exponent in the equation that describes Leonardo's hypothesis is not always equal to 2 but rather varies between 1.8 and 2.3 depending on the geometry of the specific species of tree.*

Well, we're not aiming to exactly mimic nature anyway, but it seems a good choice to make this exponent a parameter. So, the parent radius R is related to the child radii by $R^p = r_1^p + r_2^p$. Let us call p the Pythagoras exponent as it produces the Pythagoras tree when it is 2. When p is greater than two, the child branches split at a wider angle than 90°, when it is less than two they split at a more acute angle.

A tree that branches at regular intervals is simply a generalised *Pythagoras tree* at this stage, but we would like the branch border to be fractal. To enable this, we can make use of the *Cantor set*. This set is constructed from a line segment by repeatedly removing the middle-third. In the case of a branch, we can repeatedly remove one-third of the branch width (to the power of the exponent p) every half-way point along the branch segment. Removing this width is done using the junction that we have just described.

Figure 4.   Left: the horizontal lines show the reduction in branch width (to the power of p) with each junction from base to tip. This is a generalised form of the Cantor function, which is based on the *Cantor set*. Right: each iteration of the tree splits every segment in half and adds a new branch at these inserted junctions.

Removing one-third from the same side of the branch, each iteration causes the branch to curl over significantly. If we instead flip the branch direction at each split, then the tree is more balanced, and the alternating bend directions turn out to mirror those of the *Koch curve*. The one-third rule for the branch width is quite arbitrary however, so we will make this ratio a second parameter, the Cantor parameter c.

Figure 5.   Trees for $c = 0.52$. Above: $p = 2$ gives right-angled junctions. Next page: $p = 1.4$ gives obtuse-angled junctions and a much wider base. Right images make the junction triangles clear.

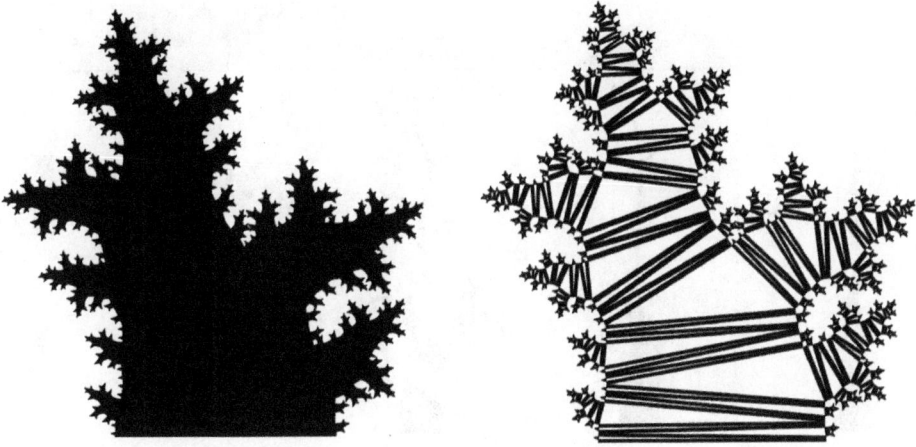

Figure 5.   *(Continued)*

We can alternatively parameterise this tree by the gradient g of its trunk, which is a function of p and c:

$$g = 4\frac{\frac{1}{2} - \sqrt[p]{\frac{1-c}{2}}}{\sqrt[p]{c}}$$

This makes the width of the tree a bit easier to control.

Plotting the results for a range of g against p shows a wide variety of trees. From short-branched, dead-looking trees and thick bracken to more fern-like or pine-like forms. The branches can even be made so short as to resemble a rough-edged rock or mountain (as shown in Figure 7).

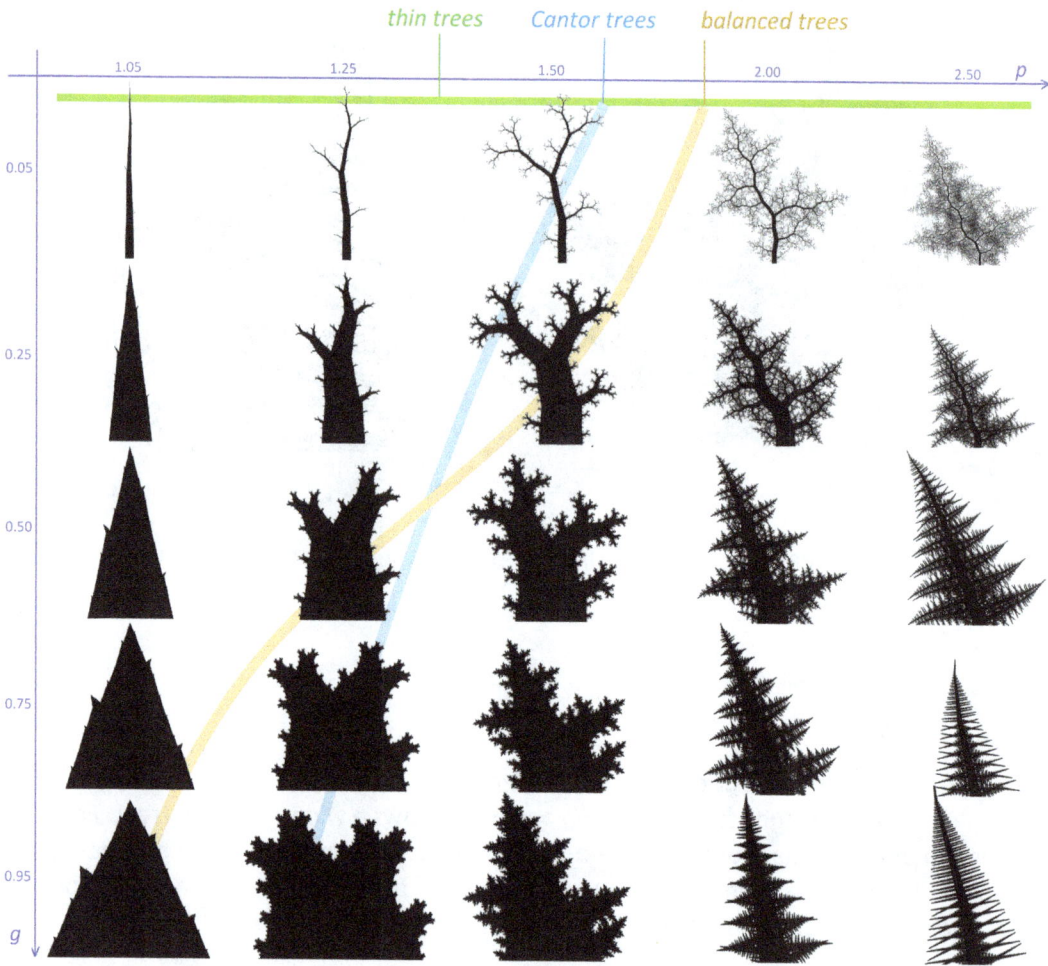

Figure 6.   Family of trees plotted by Pythagoras exponent p against gradient g, displaying a variety of forms.

Within this two-parameter family are three one-dimensional sub-families of interest. The first is the sub-family of thin trees taken as $g \to 0$. These are fractals as they have no solid 2D area. The second is the family of trees with exactly equal, Y-shaped main branch widths. These are in fact the family where $c = \frac{1}{3}$, so I will label them Cantor trees, like the *Cantor middle-third set*. The third sub-family are the balanced trees, these trees have their centre of mass directly over the middle of their trunk. Due to scale symmetry, this means the same is true of any branch that is plucked from the tree and stood up on its stem.

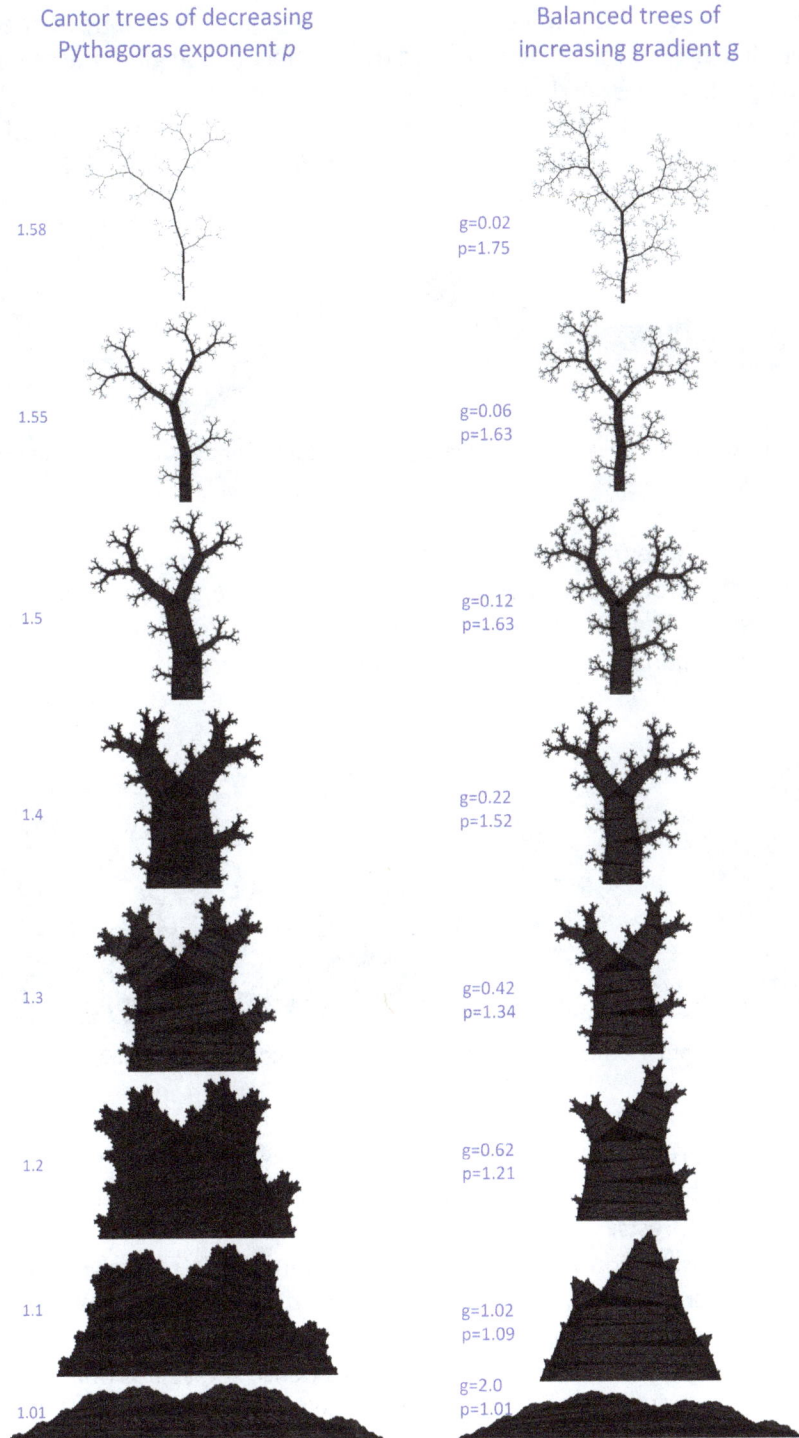

Cantor trees of decreasing Pythagoras exponent *p*

Balanced trees of increasing gradient g

1.58

1.55

1.5

1.4

1.3

1.2

1.1

1.01

g=0.02
p=1.75

g=0.06
p=1.63

g=0.12
p=1.63

g=0.22
p=1.52

g=0.42
p=1.34

g=0.62
p=1.21

g=1.02
p=1.09

g=2.0
p=1.01

Figure 7.   Balanced and Cantor trees (*c* = ⅓) plotted with respect to the remaining parameter.

It is interesting to look at just a single stem from one of these trees. It certainly has some similarity to the *Koch curve*, but the branch bend angle reduces when the section length becomes smaller compared to the branch width. As a result, the edge along this single stem is not a fractal, and its length is not infinite. It is another type of shape that is similar to, but not within the definition of a fractal. However, the border of the whole tree is indeed fractal.

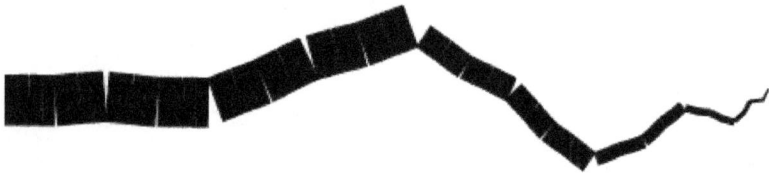

Figure 8.   The shape of a single stem matches the bend directions of the Koch curve, but with diminishing bend angle.

With any of these trees, it is interesting to zoom in on the boundary of a branch. At the limit, the branch becomes infinitely large, giving a solid half-plane that can represent ground. The offshoots become like a simple representation of a forest, and different c and p parameters give different forest densities.

Figure 9.   Zooming into the tree surface shows a forest-like structure. Left: $c = 0.52$, $p = 2$; right: $c = \frac{1}{3}$, $p = 1.55$.

Like the empirical findings of real forests [7], these forests have the same power law for the distribution of trees as for the distribution of branches per tree; they are scale-symmetric forests.

This tree construction was a quite complex example of a recursive shape. There are many simpler examples, such as the solid *Koch snowflake* and the *Pythagoras tree*. But recursive shapes aren't just tree structures.

The following example is a set of loops. It is a dense set, so it is not a fractal. Nor is its border a fractal, or its skeleton, and it doesn't have fractal leaf points. Instead, it is the skeleton's branch points that are the fractal.

The set's *multiplication rule* adds three rings of one-third the radius to each parent ring, it is a little unusual in that it also removes part of the parent ring under each child. Practically speaking, this is just achieved by rendering the child ring as an outer disk in black and then an inner disk in white.

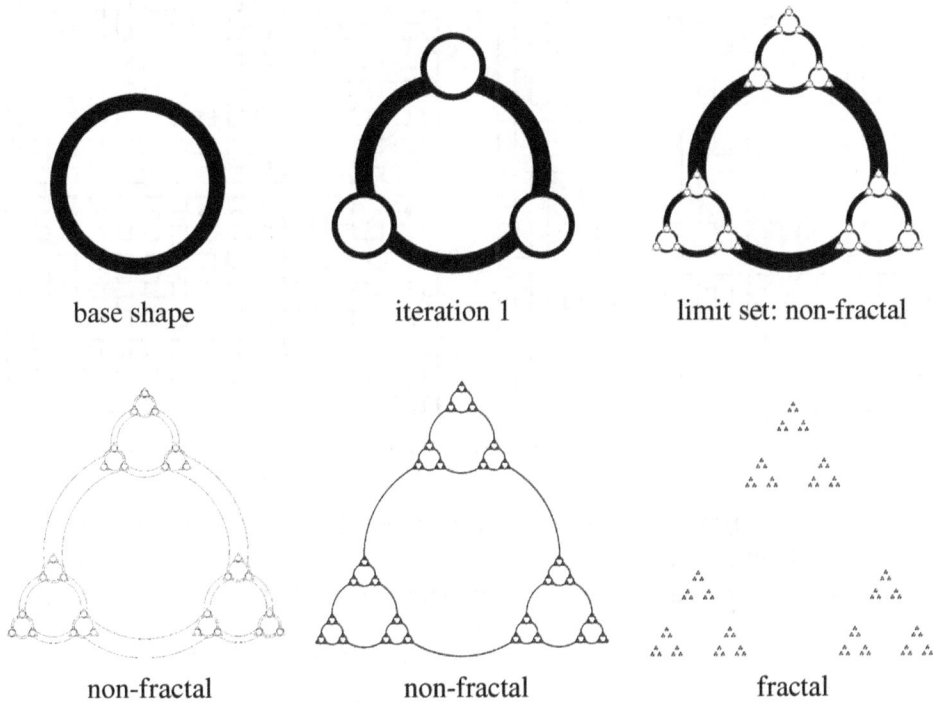

base shape	iteration 1	limit set: non-fractal

non-fractal	non-fractal	fractal

Figure 10.   Top: recursive rings structure for iterations 0, 1 and ∞. Bottom: showing the boundary, skeleton and branch points (left to right). Only the branch points are a fractal.

The skeleton of this set is itself a scale-symmetric structure and is reminiscent of connected bubbles in 2D, the circular edges are a necessary property of such bubbles, but the edge connection angles should all be 120° which they are not, so it is not particularly realistic. We will however explore more realistic bubbles in Chapter 4 on sphere inversion.

Another type of object that doesn't fit neatly within our definition of fractal is the curious case of space-filling structures. The most well-known of these are *space-filling curves*, such as the *Peano* and *Hilbert curves*.

Figure 11.   *Peano curve* (left) and *Hilbert curve* (right), with first three iterations (top).

There are also space-filling trees, we can generate a family of these by modifying the dendrite fractals from Chapter 2. The replacement scheme now removes the central shape at each iteration and just attaches *n* copies of the shape to the centre point.

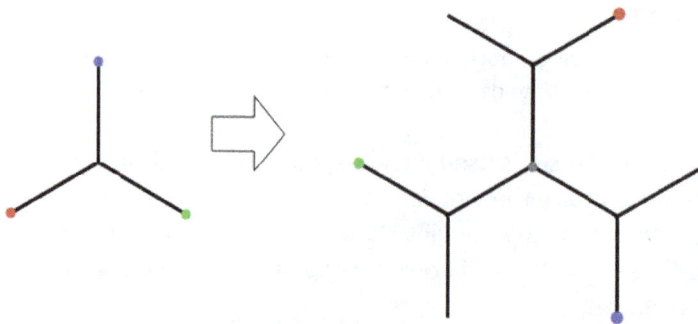

When $n = 3$, the result is a space filling tree with a fractal boundary, known as the *Fudgeflake*. When $n = 4$, we get the *Greek Cross* fractal with a simple diamond boundary. For $n = 5$ and $n = 6$, the trees overfill the 2D plane with branches that cross over each other.

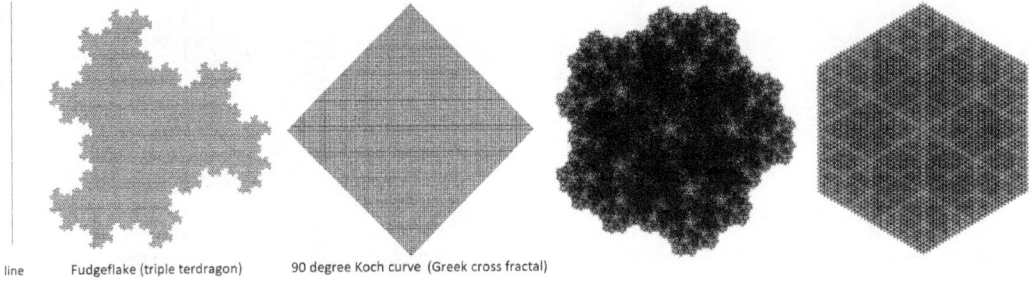

line   Fudgeflake (triple terdragon)   90 degree Koch curve (Greek cross fractal)

Figure 12.   The family of cross fractals, $n = 2$ to $n = 6$.

The amount of overfill of each pixel can be presented in grey scale. In the case of $n = 5$ in particular, this exhibits intricate patterning within the filled space.

Figure 13.   Close up of the $n = 5$ case shows pentagonal details when shaded by overlap count.

The status of these space-filling structures is a delicate matter that deserves some discussion. Taking the example of the *Peano curve*, we can define it through the curve function $f_i(s)$ at iteration i that maps an interval of the real numbers s to a subset of 2D space. If we define the Peano curve as the limit set: $L = \lim_{i \to \infty} f_i(s)$, then it is nothing more than a filled square.

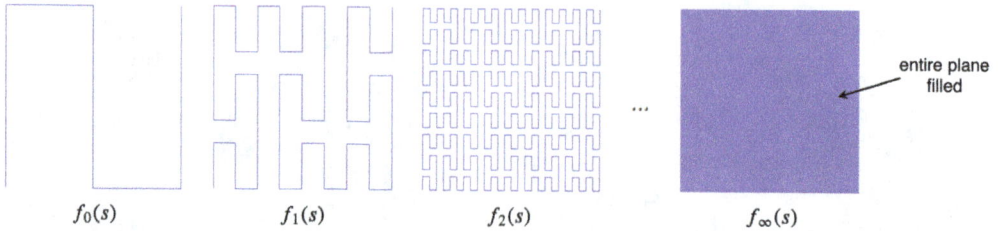

$$f_0(s) \qquad\qquad f_1(s) \qquad\qquad f_2(s) \qquad\qquad\qquad f_\infty(s)$$

Figure 14.   Peano curve as a subset of 2D space, at iterations 0, 1, 2 and at its limit, where it is no longer a curve.

Consequently, these space-filling curves refer to the function itself: $f = \lim_{i \to \infty} f_i$ rather than the result (or *image*) of the function. But functions themselves are not subsets of Euclidean space, so are not fractals by our definition.

We could take the *graph* of the function, which is the subset of 3D space $\{(s, f(s))\}$, but this is no longer space-filling. So, whichever way we look at it, the result is not a space-filling fractal and by extension nor are the other space-filling curves and trees.

The problem with taking the limit of the sequence of iterations is that the set loses its structure and becomes just a solid area. There is however a means by which the structure within these shapes can be preserved. Let us take for example a space-filling grid, this can be defined using the set $f_i = \{p : 2^i p_x \in \mathbb{N} \text{ or } 2^i p_y \in \mathbb{N}\}$, where $\mathbb{N}$ are the natural numbers. The limit set $L = \lim_{i \to \infty} f_i$ is simply the filled plane $\mathbb{R}^2$, however the set M is different:

$$M = \bigcup_{i \in \mathbb{N}} f_i$$

It is a union over the sets at all finite iterations. M has no finite sized gaps, but it is missing points, for example, $(\frac{1}{3}, \frac{1}{3})$ is missing, as are points where both coordinates are irrational. Any point in the set is part of a vertical or horizontal continuous line in the set, so it has retained its grid structure, but at this infinitely small scale.

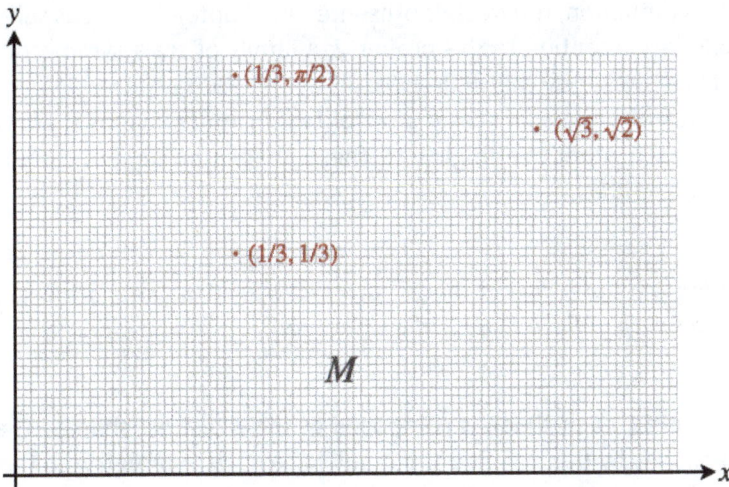

Figure 15.   Space-filling grid *M* with three examples of gaps.

I am sure that equivalent definitions for the other space-filling curves and trees are possible and they will retain their discrete scale symmetries, let's call these sets *rational space-filling* structures. However, even these curious sets would still not count as fractals. This is because fractals require the fractal (*Hausdorff*) dimension to be larger than their topological dimension, and countable unions do not increase the *Hausdorff dimension* of a set. For instance, the Hausdorff dimension of *M* is still 1, the same as its topological dimension.

So, none of these space-filling structures qualify as fractals under our definition, they are however scale symmetric just like all the previous examples. We will discuss the special status of space-filling structures again in Chapter 8, and their relationship to tilings and lattices in Chapter 9.

★

The final examples in this chapter are in fact fractals by the definition used in this book, but they are not *self-similar* in a strict sense. I include them because self-similarity is often considered to be the key property of fractals, so it is interesting to investigate the existence of scale-symmetric shapes that lack this property.

Self-similar means that parts of a shape are approximately or exactly similar to the whole shape. A stricter definition is:

Self-similarity: A shape that is composed of a finite number of smaller exact copies of that shape.

Under this definition, the fractals presented in Chapter 2 count as self-similar. So, let us consider an exception to this property: a family of scale-symmetric structures that are not self-similar by this definition.

We will look at an unusual case: a set of two or more shapes that are mutually composed of each other, we might call these shapes *co-similar*, rather than *self-similar*. We define it as:

> **Co-similarity:** *A set of n distinct shapes such that there is an arrangement of the set that produces each of the n shapes, at a larger size.*

In other words, each shape is composed of the *n* distinct shapes. These shapes are not self-similar by our definition, but still have scale symmetry, so they make an interesting edge case to explore.

Let's examine the idea more closely for the case of $n = 2$. This means there are two shapes A and B, where A is composed of a smaller copy of A and B in some arrangement, and B is composed of a smaller copy of A and B in a different arrangement.

Now that we have a definition of *co-similarity*, we can look for examples of such shapes. If we allow the child shapes to be reflections of their parents, then the problem has been studied under the name Self-Tiling Tile Sets [13], and for $n = 2$, there is a solution called the *twin-tiles*. These are simply the two principle triangles of any parallelogram, which can be placed in the two configurations by pivoting one triangle around the parallelogram centre.

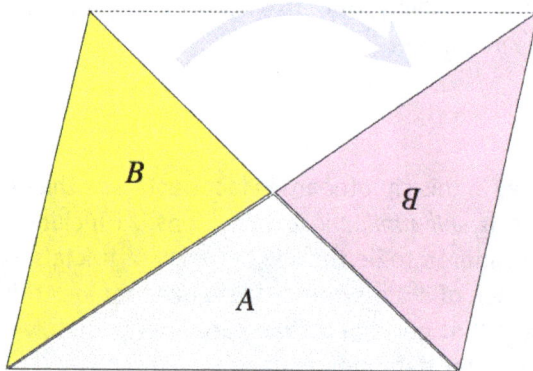

Figure 16.   *Twin-tiles*: A plus yellow B is a scaled reflection of B, A plus pink B is a scaled reflection of A.

However, if we exclude reflections, then available *co-similar* structures are far more complex. We can compute these geometries by selecting in advance what the transforms and scales of the child shapes should be within shapes A and B, then running an iterative algorithm to converge on the set of points that must make up such a structure.

Figure 17.   Co-similar shapes. White is shape *A*, yellow or pink is shape *B*. white + yellow = shape *A*, white + pink = shape *B*.

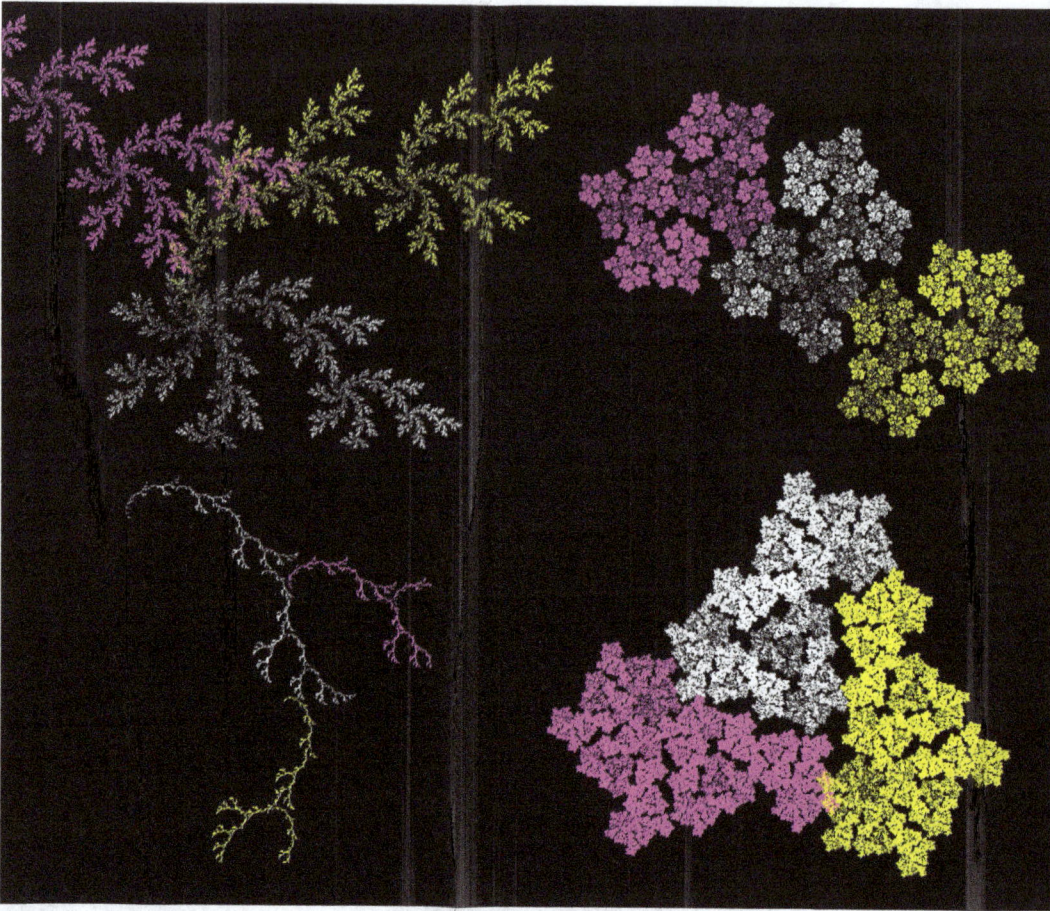

Figure 17. (*Continued*)

The results are striking, while they have a similarity with traditional limit sets, the variety within the shape seems to be greater. For instance, some examples show spirals in both directions. Other than spirals, there are also shapes similar to plants, leaves, wreaths, lightning, and feathers.

Construction

Keeping shape A at identity, we can define the two arrangements of B by the two Euclidean transformations T_A and T_B, which create a larger transformed version of A and B defined by the Euclidean transforms V_A and V_B, respectively, and the scale factors s_A and s_B.

The problem is then to find set A and B given a transform set T_A, T_B, V_A, V_B, s_A and s_B such that

$$T_A B \cup A = s_A V_A A \text{ and } T_B B \cup A = s_B V_B B$$

This is found by rearranging the above into a recurrence relation:

$$A_{i+1} = s_A^{-1} V_A^{-1} \left(T_A B_i \cup A_i \right)$$

$$B_{i+1} = s_B^{-1} V_B^{-1} \left(T_B B_i \cup A_i \right)$$

which can be initialised with $A_0, B_0 = \{(0,0)\}$. The limit as $i \to \infty$ gives the two co-similar shapes.

These *co-similar* sets are in general disconnected, but the ones shown here are filtered to just those that are connected or close to connected, while not overlapping. Although connected structures exist, it remains an open question whether there is a structure that is dense in the 2D plane. We can get arbitrarily close to a pair of dense tiles A and B by converging on self-similar shapes, such as a 45° right-angled triangle or a $1 \times \sqrt{2}$ rectangle, but it doesn't make for an interesting example. If the 2D shape is allowed to exist in 3D, then the *twin-tile* becomes a solution because a reflection can be achieved just by turning the plane over, which is a 3D rotation. But in two dimensions the closest to a dense pair of tiles that I have found is the final image of Figure 17. The total search space is large, but it seems more likely than not that all two-dimensional co-dimensional shapes are not dense. These non-dense structures qualify as fractals, but they are not self-similar by our definition.

★

As you can see from the above examples, there are a great variety of scale-symmetric structures that are not fractal. Fractals are just one special case of scale-symmetric geometry, but the subject in general is much broader. The following chapters explore the many forms of this general-case scale symmetry. These are just as beautiful and are even more diverse and remarkable than fractals alone.

Chapter 4

Sphere Inversion

In the introduction, we defined scale-symmetric shapes as those that *look the same under shape-preserving transformations that include scaling,* where the basic shape-preserving transformations are translations, reflections, rotations and uniform scaling. The examples up until now have used combinations of these four uniform transformations. However, there is one other transformation that is shape-preserving in every number of dimensions, and this is the one we shall explore in this chapter. It is called *sphere inversion.*

We defined *shape-preserving* in the introduction as transformations that pre-serve the shape of small features. We can be a bit more specific now:

> **Shape-preserving:** *A transformation under which small spheres remain spheres. Where 'small' is used to mean the limit as the radius tends to 0.*

The characteristic of shape-preserving transformations is that they never squash or stretch features unevenly, and the angles at corners are always maintained. We have used the word *sphere* here as shorthand for any *n*-dimensional sphere, such as a circle in 2D or a standard sphere in 3D. From this definition, it should be clear that rotations, translations, reflections and uniform scalings all transform small spheres into spheres, in fact they transform any sized spheres into spheres. It should also be clear that corners on a grid remain right-angled after applying these transformations.

The *sphere inversion* transformation shares these same properties. It is a simple operation: one simply replaces the length of every vector in the space with its reciprocal length. Mathematically, $p \leftarrow \frac{p}{|p|^2}$. So, it inverts vectors about the sphere; points inside the sphere become outside and vice versa.

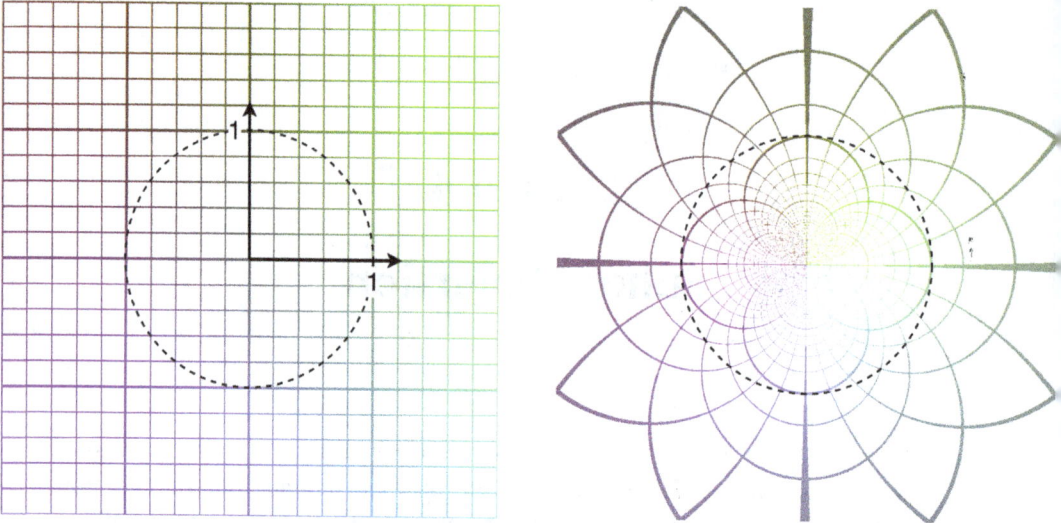

Figure 1.   Left: grid of points p. Right: the sphere inversion of this grid is simply a radial reciprocal ($p/|p|^2$), which "reflects" the grid around the dashed unit circle (or sphere in 3D). Note that all corners remain right angles when transformed.

If we combine *sphere inversion* with one reflection and combine it with the other orthogonal transformations (translation, rotation and scale), then the general name for these is *Mobius transformations*. We will discuss these and more general transforms in Chapter 5, but this chapter is dedicated to *sphere inversion*. Since sphere inversion always transforms spheres into spheres, it follows that recursively applying it to spheres will give self-similar results, as the spheres will never become flattened or elongated.

This makes it a useful transform for a sort of structure that we have so far only mentioned in the introduction. That is, those built from separate bodies in a scale-symmetric fashion, so that there are exponentially more bodies with diminishing radius. This is well represented by structures in space such as the asteroid belt, Saturn's rings, galaxies and galaxy clusters, let us call this structure a *cluster*.

Figure 2.   *Clusters*: the approximate scale-symmetric distribution of different sized objects. Left: asteroids and space debris. Right: water droplets on a window.

Under gravity, clusters coalesce into a big conglomeration of contacting bodies, an example of a self-contacting structure like this is an asteroid, where the bodies are the boulders, rubble, and dust that it is made of:

Figure 3.   *Clusters* under gravity. Images of the surface of asteroid *162173 Ryugu* by the Hayabusa2 sample return mission. The surface shows several large rocks and increasing numbers of smaller rocks.

On smooth surfaces, the smallest bodies settle at the bottom, this has the effect of hiding the range of scales and masking the scale symmetry, so even uniform looking structures such as pebble beaches can be scale-symmetric as well.

Figure 4.   Pebbles on a beach appear to have a limited size range, but smaller pebbles often sit underneath, so the entire structure can still have scale symmetry.

Let's construct such a cluster — a mathematical pebble beach — using spheres on a planar surface. We will densely pack the plane with spheres so that all spheres touch the plane and contact each other, this is sometimes called a *Ford sphere* packing. We start with a dense packing of spheres of constant radius, giving a triangular lattice of spheres. The construction is then conceptually simple:

For every triplet of contacting spheres, we can pack in a fourth, smaller sphere that contacts all three and the plane, this fourth sphere creates three new triplets with which to repeat the process.

This is a sufficient description to generate the entire structure, which is a maximal packing of spheres onto the plane.

Figure 5. Close-up of *Ford Sphere set*, several smaller spheres are visible through the gap, but are mainly hidden by the largest size spheres, which are packed in a triangular lattice.

However, the structure can be calculated more directly, in a surprisingly succinct formula, as given in the panel below. The result is rendered as a long list of spheres.

Ford Sphere Set — Definition

Using the complex plane as the horizontal plane, the set is:

$$\bigcup_{p,q\in\mathbb{E}}\left\{x:\left|\left(\frac{p}{q},\frac{1}{2|q|^2}\right)-x\right|<\frac{1}{2|q|^2}\right\}$$

where $\mathbb{E}$ are the Eisenstein integers, which are integer combinations of the cube roots of -1, forming a triangular lattice, and $|x|$ is the Euclidean length of the vector in the $\mathbb{C}\times\mathbb{R}$ space. More information on this formula can be found at [14].

This formula highlights that the *Ford Sphere set* is a nice representation of the Eisenstein rationals (the numbers a/b where a, b are Eisenstein integers). Large spheres are at points in the complex plane that are "simple" rationals, meaning that the numerator and denominator are small numbers. The smallest spheres require much larger numbers in the numerator and denominator to describe their location.

The neighbouring spheres of a large sphere are wedged between it and the ground plane, so must be small. This gives a direct geometric explanation for why you can't approximate "simple" Eisenstein rationals, like ½, by other simple Eisenstein rationals.

This is the traditional method for rendering scale-symmetric structures, whereby the structure is decomposed into a large list of Euclidean shapes that are fed into traditional rendering software, we have employed this method in all the previous examples. However, there is an entirely different way of rendering these structures that can be significantly more efficient. This alternative "backwards" method requires a function that maps any point in 3D space to a single value that estimates the distance of that point to the structure's surface. Specialist rendering software[1] then samples this function at many points in order to ray-trace the structure and light it appropriately.

This distance estimation technique is powerful and popular with 3D fractals. It avoids the storage requirements of the traditional "forwards" rendering and only processes what needs to be seen, so deep zooms and high detail levels are feasible. The method is used in many subsequent 3D examples, so let's look into how it works.

Fractal *distance estimation functions* all follow a similar algorithm. They treat the position argument to the function as the location of a small sphere of space defined by its centre p and its initial scale $s = 1$. They then iteratively adjust the sphere in a manner that is akin to zooming in on the nearest surface to p, and they scale s accordingly. After n iterations, the nearest surface is approximated as a simple primitive, such as a line or sphere, and the function returns the distance to the primitive, divided by s.

---

[1] Examples include Fragmentarium and Mandelbulb3D. See the back of the book for a full list of software.

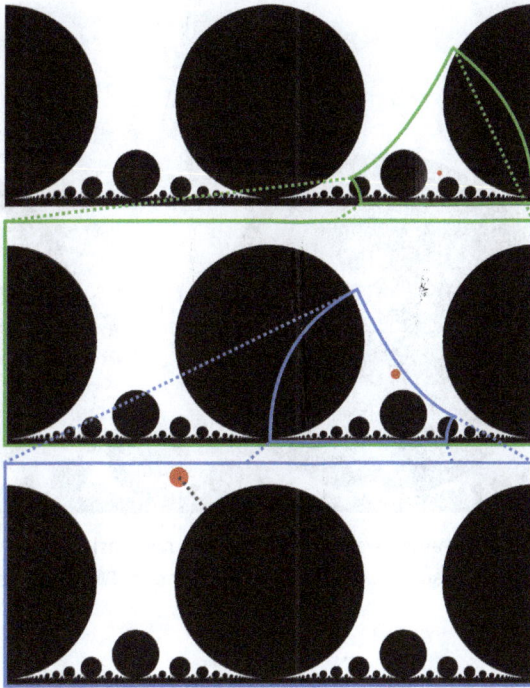

Figure 6.   Example of *distance estimation function* iterations. The input point is treated as a small radius disk (red, top) of scale $s = 1$. The iterations (middle and bottom) act to "zoom in" to the nearest self-similar part of the structure, until the distance can be approximated as the distance to the central disk, divided by the red disk's new scale s.

The rendering software then uses the estimations of the distance to the fractal surface to sample along light rays until it gets within a small distance ϵ to the surface. The user can adjust this tolerance ϵ and control the lighting and surface properties to tune the exact appearance of the 3D shape.

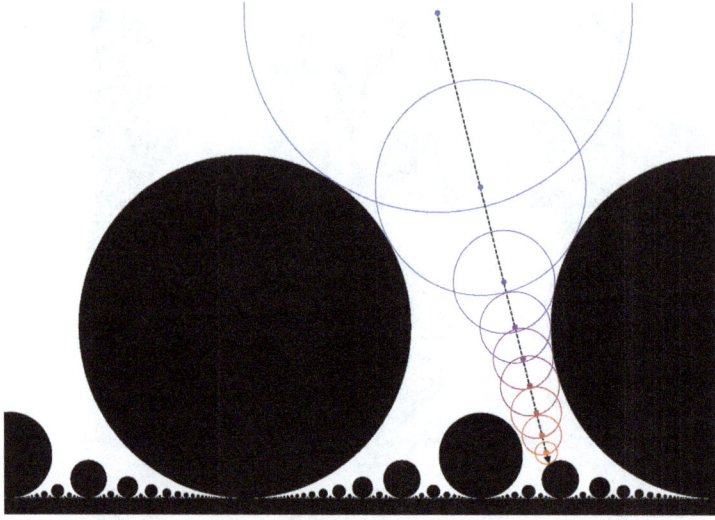

Figure 7.   Rendering software samples along each camera ray (dashed line) by a percentage of the distance estimate to the nearest surface. In this example, it moves 90% of the distance every step.

Ford Sphere Set — Distance Estimation Function

Input: 3D position p
Output: distance to surface d

Set L is a triangular lattice: integer combinations of $(0,1,0)$ and $\left(\frac{\sqrt{3}}{2}, -\frac{1}{2}, 0\right)$

– Translate p to be relative to the nearest point on L.
– Set scale $s = 1$ and loop for n iterations:

1. Viewing from the side: if p is in region A (or B on first iteration), then exit loop:

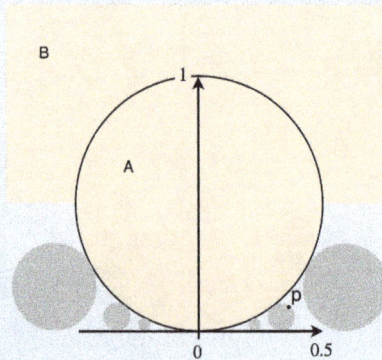

2. Otherwise we apply sphere inversion: scale p and s by $1/|p|^2$.

Viewing from above, this acts like a reflection around the black circle, and transforms the tiny spheres under the main sphere into a regular lattice of spheres:

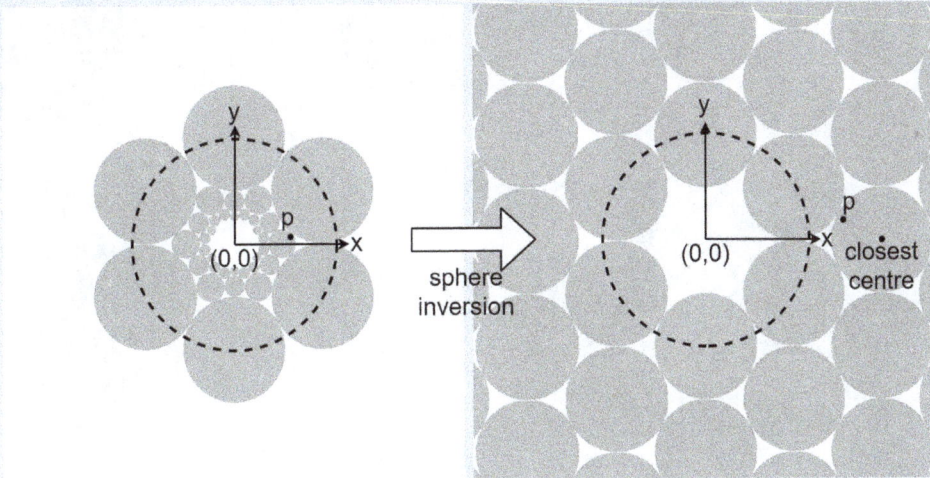

3. We now subtract the closest lattice point in L (grey sphere centre) from p:

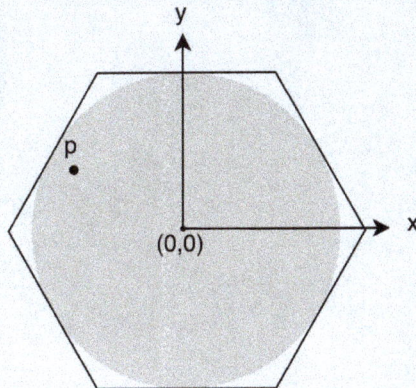

End loop.

We now use the distance to the final central sphere:

Output: $d = (|p - (0,0,\frac{1}{2})| - \frac{1}{2})/s$

To render smaller spheres, we can scale them by a constant $0 < k < 1$:

Output: $d = (|p - (0,0,k/2)| - k/2)/s$

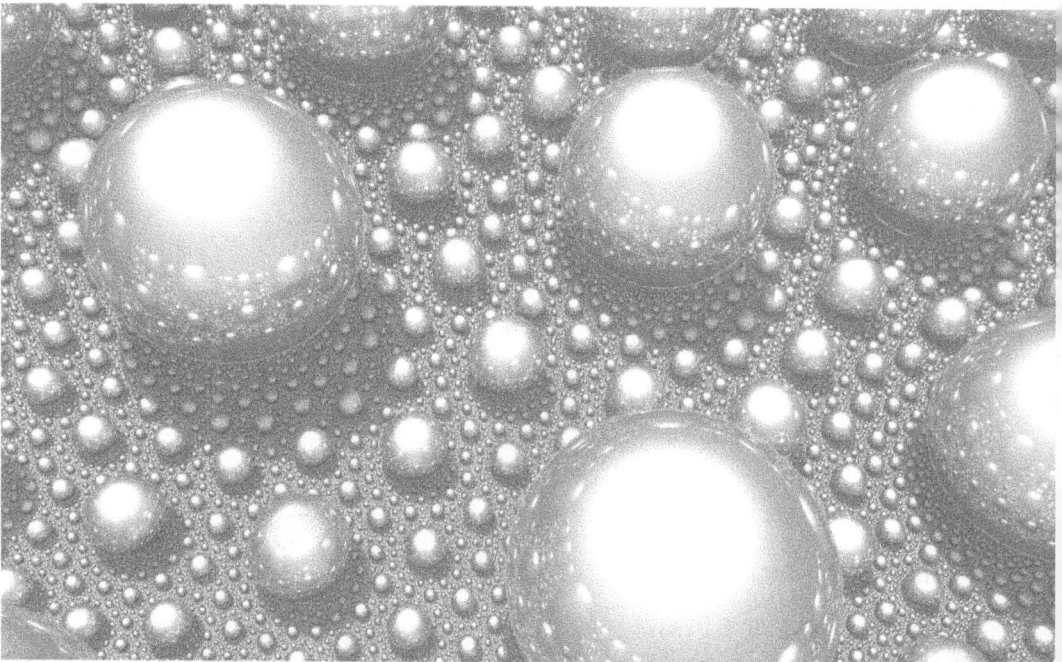

Figure 8.   *Ford Sphere set* with reduced sphere size to reveal the structure. Top: at 80% of contacting radius. Bottom: at 50% of contacting radius ($k = 0.5$).

Like the beach pebbles, the structure of spheres is obscured by the largest ones, but by downscaling the spheres their complex distribution is revealed. Note that the largest spheres only have tiny spheres near to them, this is why there is no intersection in the full scale $k = 1$ case.

Sphere inversion needn't always produce structures built of spheres, but it is well suited to structures built from sphere segments, that is, segments of constant curvature. A great example of this is connected bubbles. Let us look at the 2D case to begin with. We know that such bubbles must have edges of constant curvature and must meet at 120° angles, this is because 2D bubble dynamics minimises edge length subject to enclosing a given area per bubble. These constraints are two-dimensional equivalents of *Plateau's laws* for soap films.[2]

We can now generate a more realistic version of the bubble structure in Chapter 3, this time using sphere inversion, and this is most readily achieved with the "backwards" distance estimation technique used previously.

2D Bubbles — Distance Estimation Function

Set scale $s = 1$.
Loop:
1. Find closest vertex v to p on triangle of unit edge length:

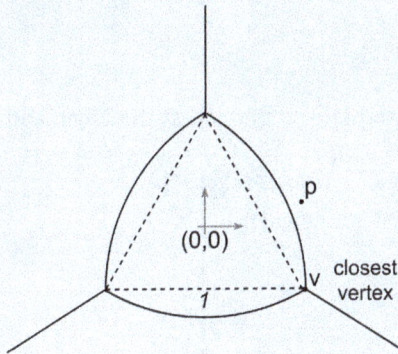

---

[2] Formulated in the 19th century by physicist Joseph Plateau, they state that soap films are a continuous surface of constant mean curvature, where faces meet in threes at 120° and edges meet in fours at $\operatorname{acos}(-1/3)°$.

2. Translate p and v such that v is radially length $l = 3$ from the origin:

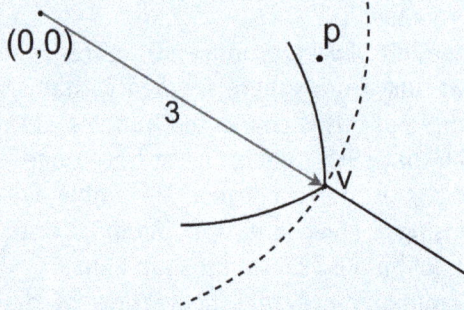

(0,0)

p

3

v

3. Sphere inversion around radius l, so scale p and s by $l^2/|p|^2$:

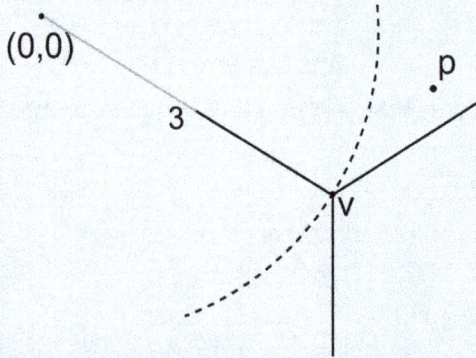

(0,0)

p

3

v

4. Reflect in v direction, translate so that v is at the origin and scale p and s by k:

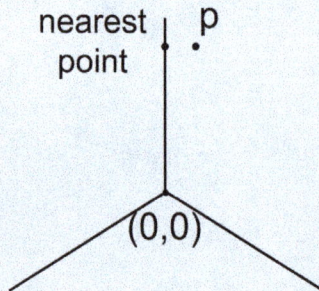

nearest point   p

(0,0)

End loop.

Output: d is the distance of p to the nearest point on the radial triangle shown above, divided by s.

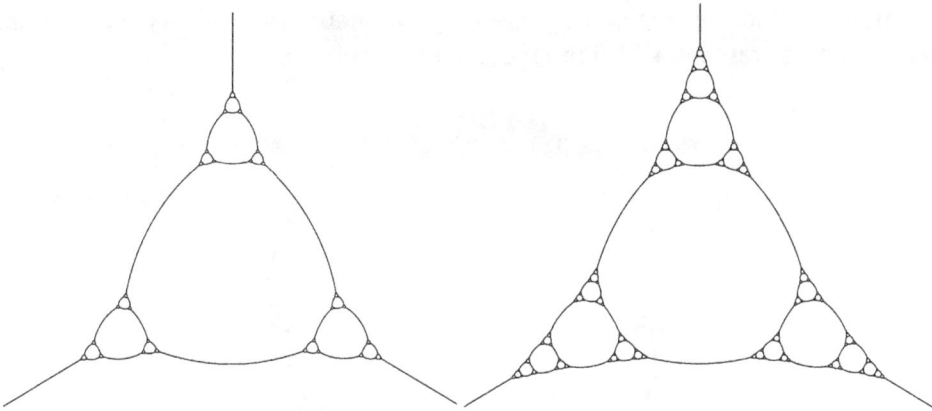

Figure 9.   2D bubbles fractal. Left: scale factor $k = 4$. Right: $k = 3$ gives larger child bubble sizes.

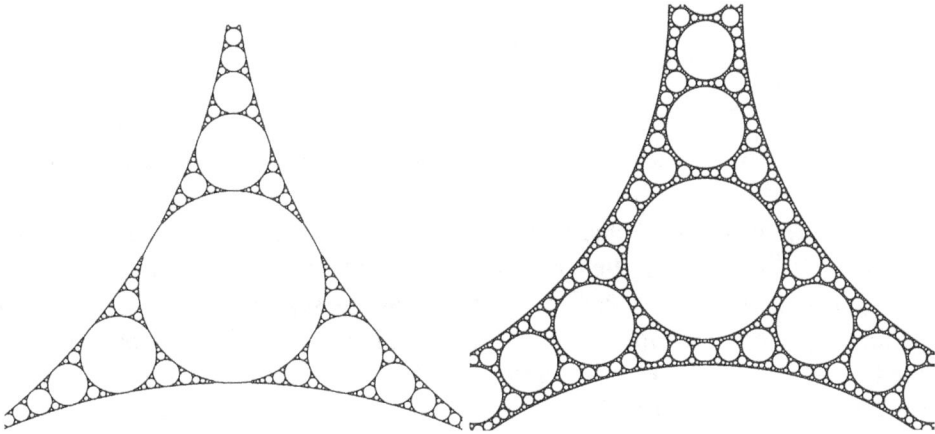

Figure 10.   2D bubbles. Left: for scale factor $k = 2.5$, the result is an *Apollonian gasket*. For lower scale factors, such as $k = 2$ (right), the results resemble various circle packings.

The results for $k \geq 3$ do appear more real than the previous chapter's bubble structure, this is due to the 120° junction constraint. For smaller k values, the child bubble sizes are larger and for $k < 2.5$, the resulting shapes appear like circle packings. At approximately $k = 2.5$, the fractal is built of full circles. This is called an *Apollonian Gasket* fractal because it iteratively solves the *Apollonian problem* of finding the circle that is tangent to its three neighbours. The word *Apollonian* here refers to the Greek mathematician Apollonius who studied the problem, and *gasket* refers to a set of circles.

A nice feature of the distance estimate method is that the final shape that we estimate the distance against can be easily modified. This changes the structure's appearance without changing its underlying symmetry. Here, we change the line segment into a crescent, which produces a rather delightful design:

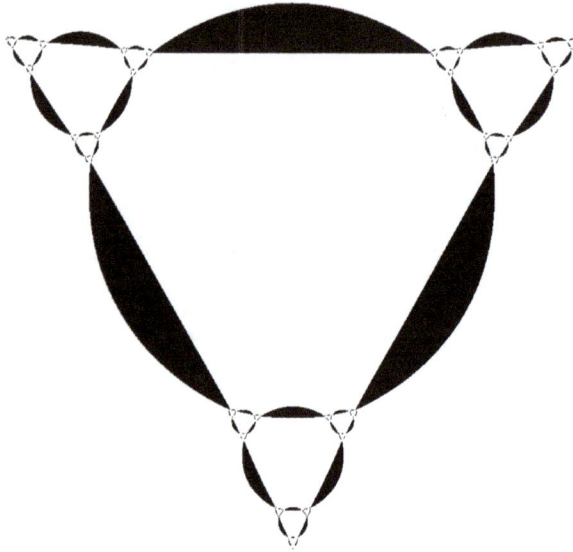

Figure 11.   2D bubble fractal, outputting the distance to a crescent instead of a line. $k = 4$.

In three dimensions, bubble surfaces also meet at 120° along edges, and these edges meet as though from the corners to the centre of a tetrahedron. The surfaces are not always of constant curvature like a sphere segment, they are generally described by *minimal surfaces*. However, non-spherical bubble surfaces arise only when there are fixed boundary conditions, such as when soap film attaches to a wire frame. When unconstrained, the surfaces are free to find the most minimal area for the fixed volume of air they enclose, this gives rise to constant curvature geometry.

Figure 12.   Left: bubbles constrained to a boundary are usually not spherical surfaces. Right: when unconstrained, each segment is spherical.

We can therefore generalise the 2D bubble structure quite directly into 3D. It is now built of curved tetrahedral structures rather than curved triangular structures.

Figure 13.   3D bubbles, rendered opaque and with reflection. Much of the structure is concealed.

Figure 14.   3D bubbles rendered with transparency and reflectivity. The internal structure is more apparent.

Figure 14.   (*Continued*)

3D Bubbles — Distance Estimation Function

Apply the 2D Bubbles function, but use the four vertices of a tetrahedron in place of the three triangle vertices. The sphere inversion now occurs at length $l = 2⅔$.

Define a *radial tetrahedron* as the set of four faces connecting each pair of vertices in the tetrahedron with the centre.

Output: d is the final distance of p to the surface of the *radial tetrahedron*, divided by s.

Foam-like structures such as this are hard to visualise and so are not commonly seen in geometry texts. However, in this case, we can emulate the transparency and reflection of real bubbles and the result is quite effective and allows the internal structure to be seen fairly well. In real bubbles, the thinness of the film produces diffraction effects that are wavelength-dependent, which result in some wonderful oily and iridescent colours on the surfaces. It would be a fun exercise to replicate this in the distance estimation rendering pipeline and so add an extra degree of realism.

We can also render just the edges by modifying the distance function to report the distance to the nearest radial tetrahedron edge rather than face. This produces sponge-like structures rather than foams.

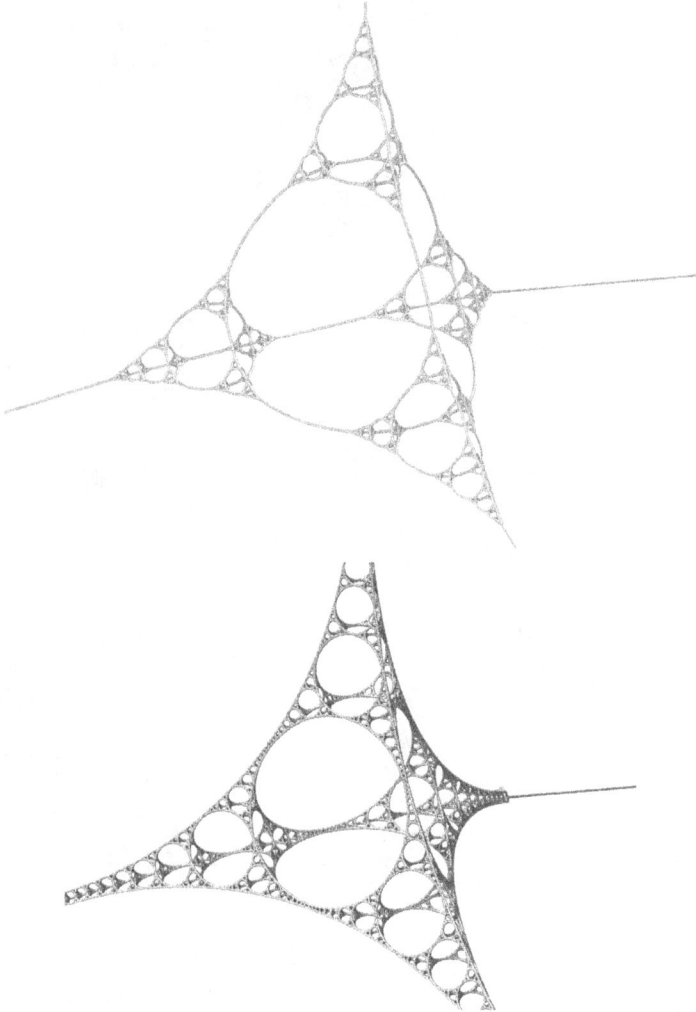

Figure 15.   3D bubbles fractal with only edges rendered shown for decreasing scale factor k.

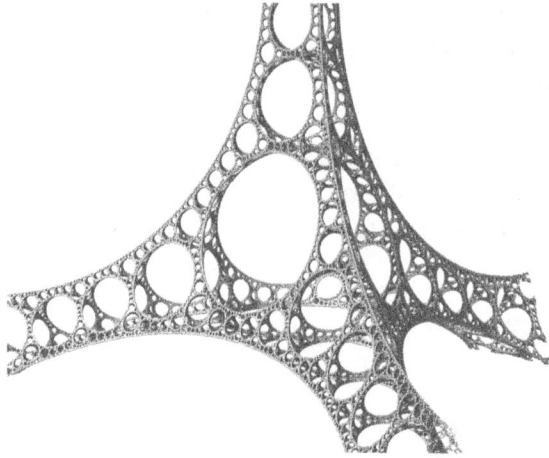

Figure 15.   (*Continued*)

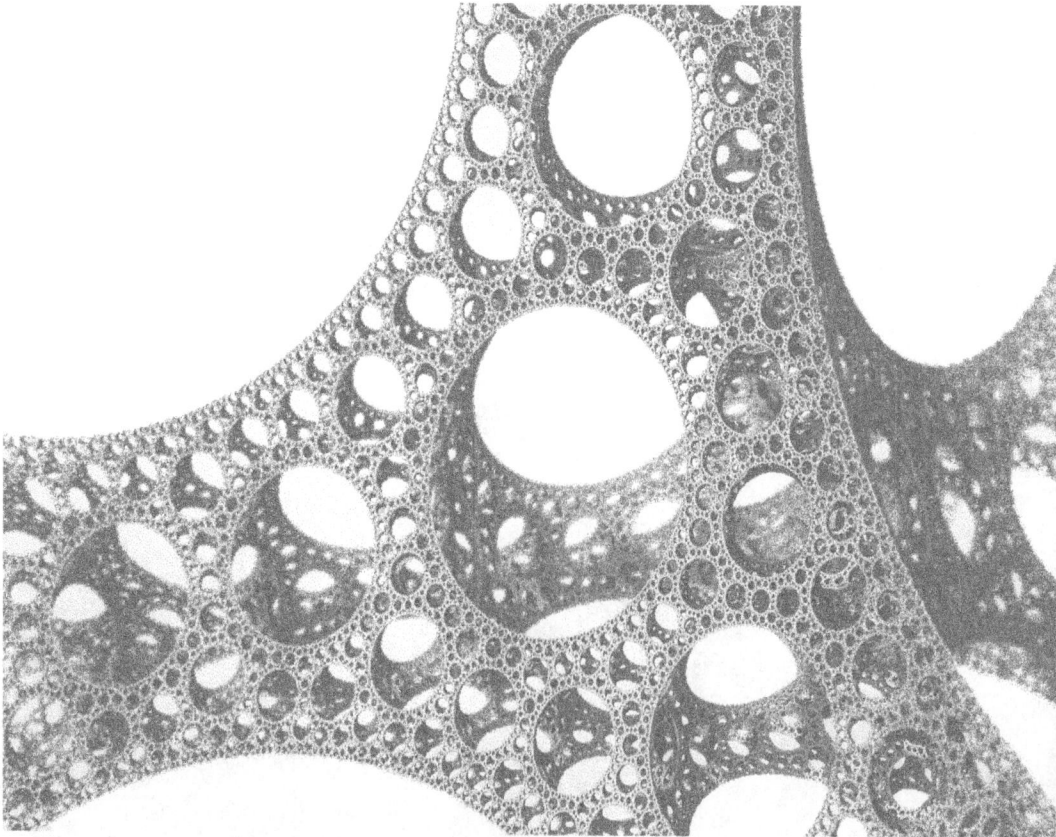

Figure 16.   3D bubbles fractal with only edges rendered. For low *k* values such as the above image, all edges are circular, producing what are called *Kleinian fractals*.

At smaller scale factors, they take on an increasingly elaborate structure of circles. These are often called *Kleinian fractals* because the set of transformations in the iteration loop forms a discrete group called a *Kleinian group*. The book *Indra's Pearls: The Vision of Felix Klein* [15] gives an excellent introduction to this topic.

These above structures have no surfaces and are a connected network of edges built of loops. They are sometimes described as sponges or sponge-like, as is the case with the 3D *Menger sponge*. So, foam-like and sponge-like shapes can be produced with sphere inversions. Are there any other sort of shapes that can be produced? The answer is yes, we can also use spheres to build tree-like shapes.

Generating a 3D tree with a fractal surface is difficult from a mathematical and rendering point of view, but sphere inversions based on the previous *Ford Sphere set* provide a good starting point, and the distance estimate technique is an efficient way to render the structure realistically. Conceptually, we can build such a tree by transforming the *Ford Sphere set* such that its upper bounding plane wraps over a large sphere, and its scale tends to zero at the sphere base. This creates a new shape that contacts the ground at a single point. The next step is to repeatedly replace all the spheres on this shape with small copies of the shape itself. There is no overlap because each iteration transforms the spheres into a smaller shape, and there are no disconnected parts because the replacement continues to sit flush against its parent surface.

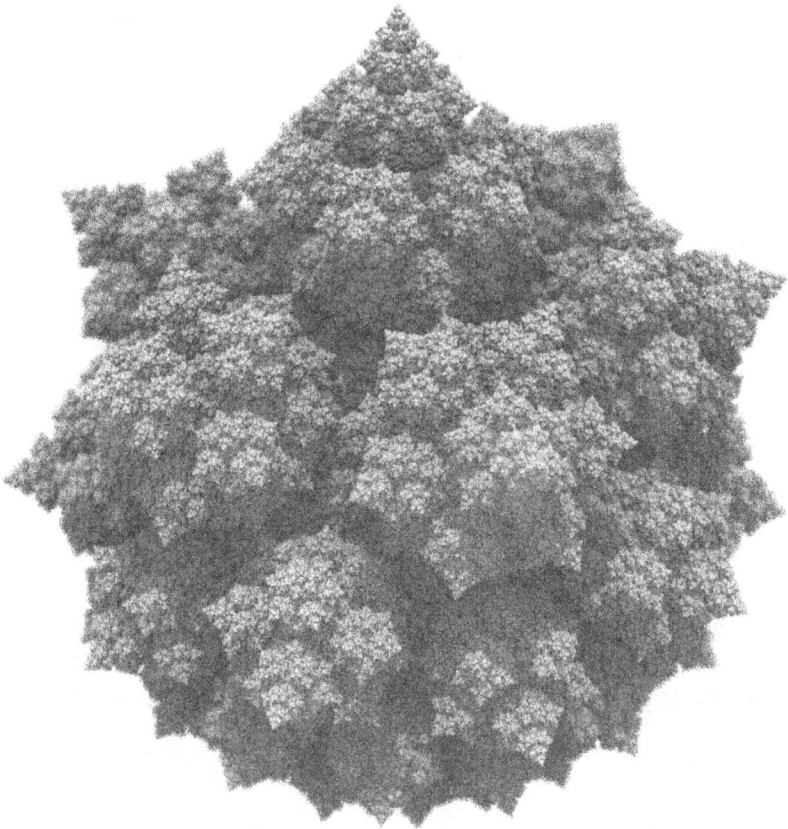

Figure 17.   Sphere tree, built from recursive Ford Sphere packings.

Sphere Tree — Distance Estimation Function

We define the shape with base at the origin and unit height. The iterative method follows steps 1–3 of the earlier *Ford Spheres* set, but we substitute step 1:

1. Viewing from the side: if p is in region A (or in region B on first iteration), then exit loop:

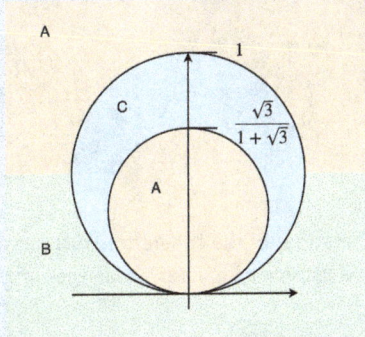

Otherwise, if p is in region C, then

a: Sphere inversion: scale p and s by $1/|p|^2$.

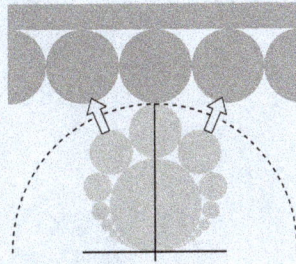

b: reflect vertically, translate to the origin and scale to unit diameter:

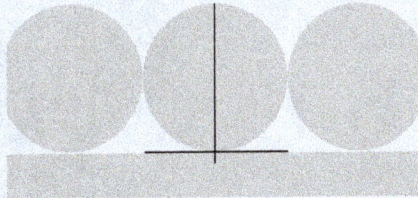

c: rotate 30° around the vertical axis.
d. go to *Ford Spheres* step 3.

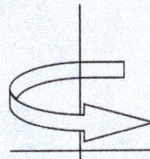

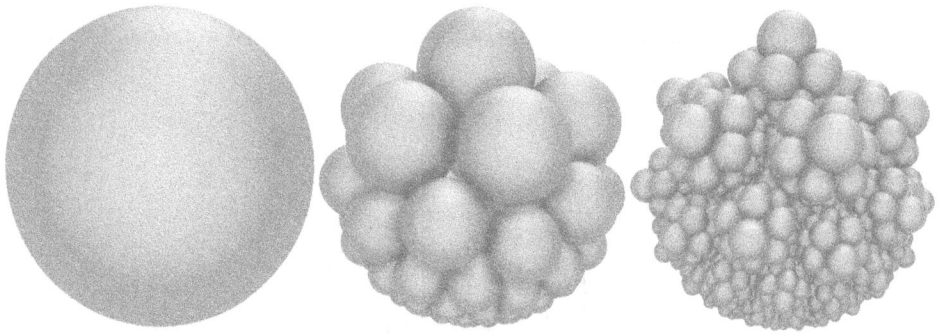

Figure 18.   Showing 0, 1 and 2 iterations of the sphere tree distance estimation function. Each iteration substitutes each sphere with a packing of spheres within the same spherical bounds, so it doesn't introduce any overlaps.

Figure 19.   Close-up of sphere tree, using the depth-of-field camera effect to help emphasise its 3D structure.

It might be tempting to think of this as some sort of 3D analogue of the famous *Mandelbrot set,* but it isn't. This is its own unique shape and is in fact a 3D analogue of the following configuration of recursive *Ford circles*:

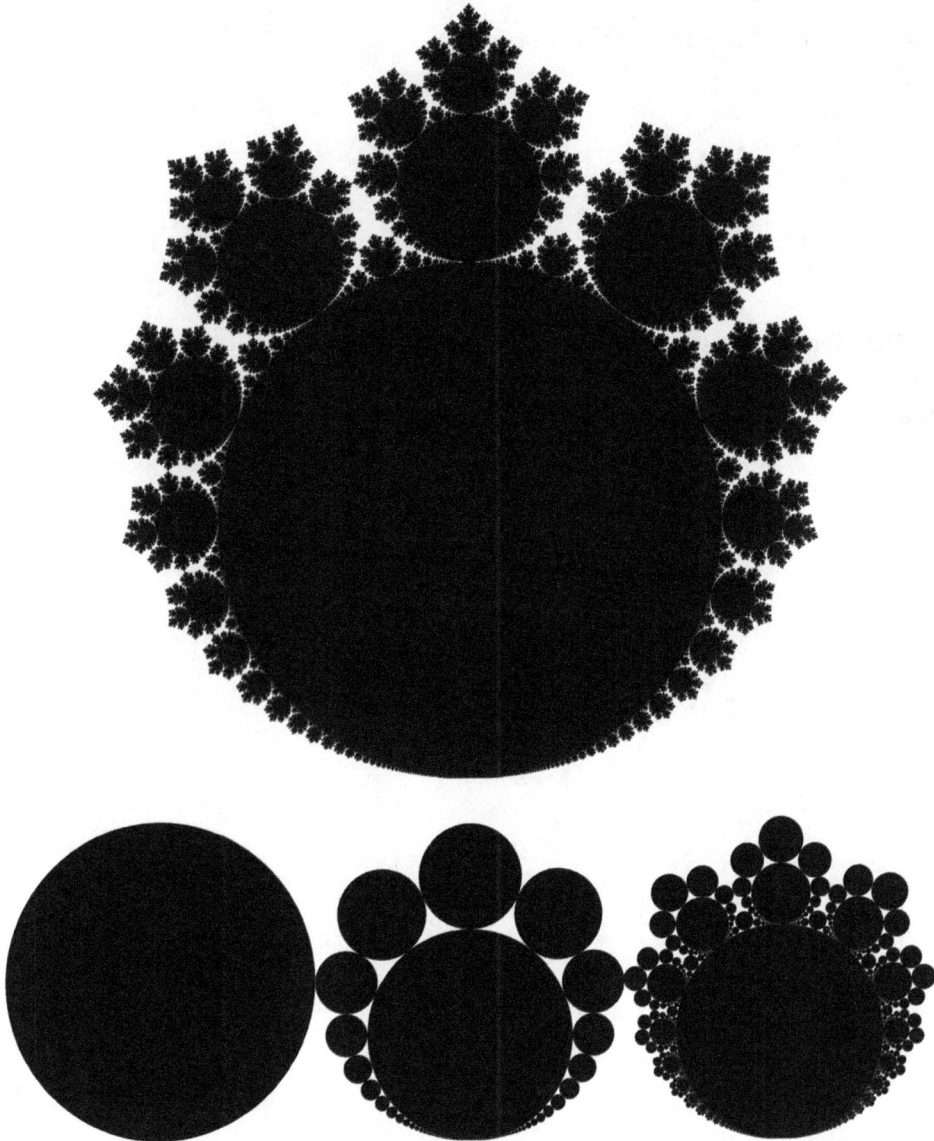

Figure 20. Top: 2D analogue of prior 3D recursive spheres. Bottom: the first three iterations.

2D Tree — Inside Set Function

Input: position p

while $|p - (0,1/3)| > 1/3$ and iterations $< n$:

$p \leftarrow p/|p|^2$

if $p_y > 1$:

$p = (0,3) - 2p$

$p_x = [p_x]$ where [] means round to the nearest integer.

Inside set is: $|p - (0,1/2)| < 1/2$.

The Mandelbrot set is an altogether different beast and uses a different type of shape-preserving transform than sphere inversion. We will talk about this wonderful set and its unusual properties, and explore its generalisations in the next chapter.

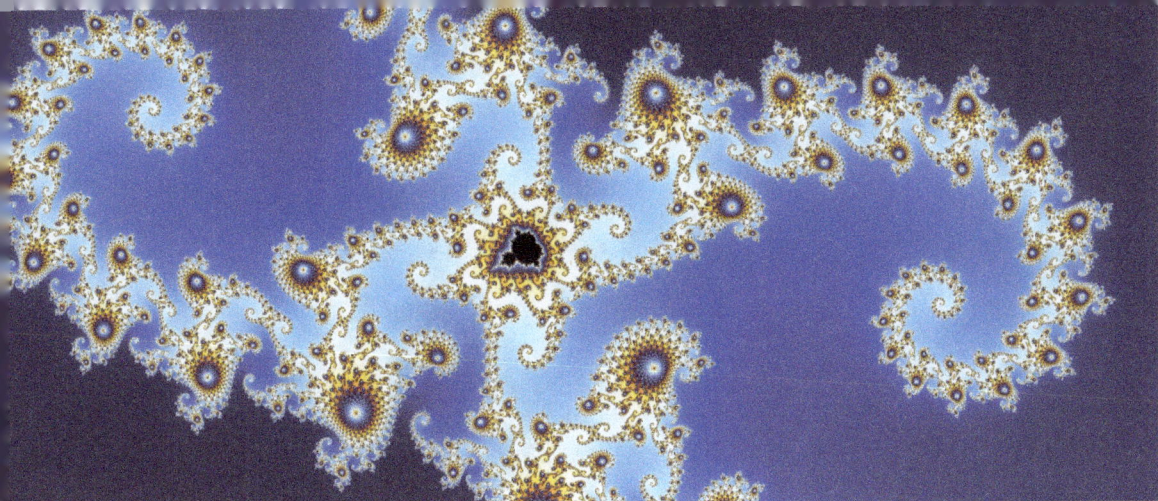

Chapter 5

Mandelbrot Sets

The Mandelbrot set is the set of complex numbers z_0 that have a bounded iteration of the function:

$$z_{i+1} = z_i^2 + z_0$$

So, when rendering, if the magnitude of z_i grows without bound after n iterations, then the pixel at z_0 is not in the set and is usually coloured as a function of n. All remaining black pixels represent *the Mandelbrot set* itself.

Geometrically, the complex squaring operation doubles the angle of z from the positive real line and squares its magnitude.

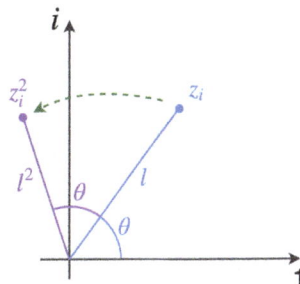

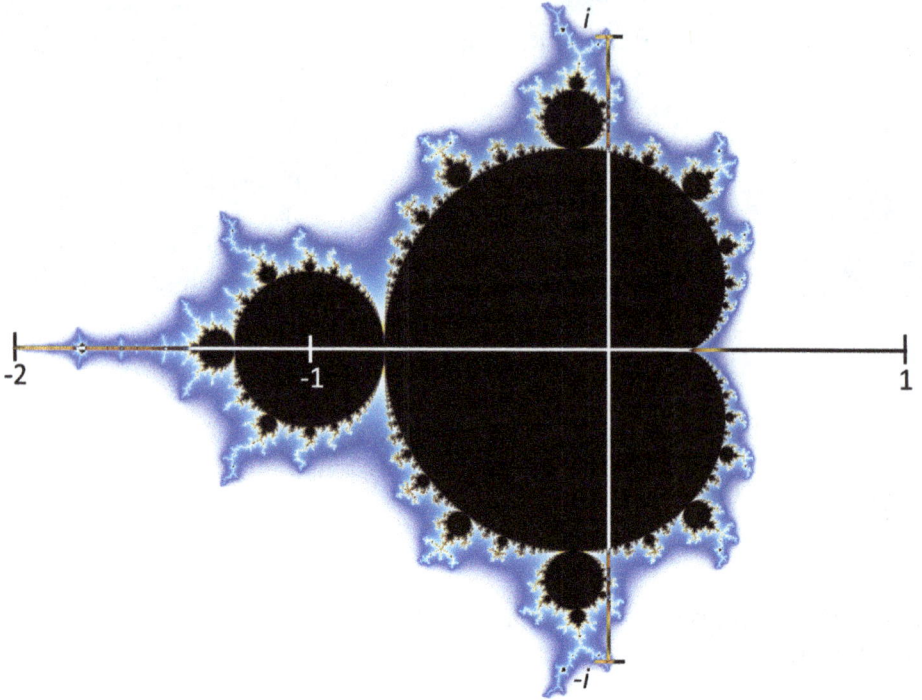

Figure 1.   The Mandelbrot set plotted on the complex plane. Black points are in the set, coloured points follow a gradient that indicates proximity to the set.

So, the operation on the whole complex plane $\mathbb{C}$ acts to wrap it around itself:

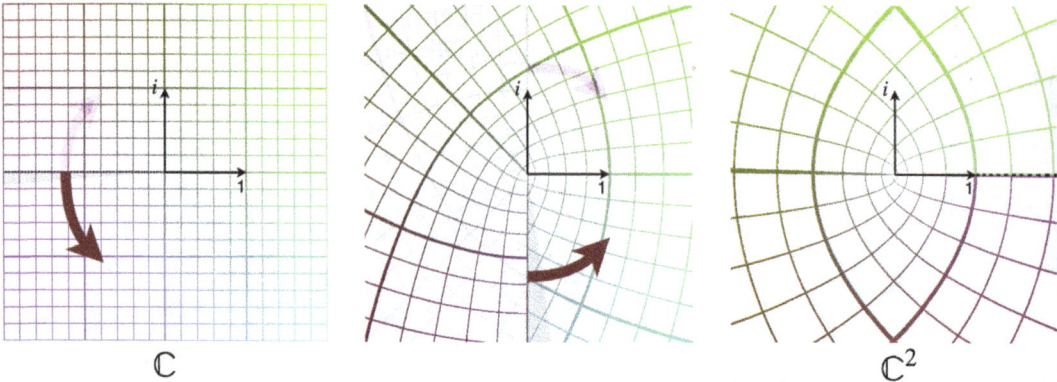

Figure 2.   The z^2 operation on the whole complex plane is equivalent to cutting along the negative number line and pulling these edges around to meet along the positive number line.

This produces a double-cover of the plane, meaning that two values of z lead to each z^2. This square operation also preserves angles, in the above image, you can see that every warped square in the grid still has 90° corners. This means that small features on the complex plane are not squashed or elongated by the squaring operation, it is a *conformal transformation*.

Conformal transformations include translations, rotations and uniform scalings, and the sphere inversion operation of Chapter 4 when combined with a reflection. In the complex plane, they also include all complex power functions z^n. They are at the heart of scale-symmetric geometry as they are the most general type of *shape-preserving transformation* of those that preserve handedness. This handedness preservation means that a small letter **b** always transforms to a **b** and never to a **d**.

The simpler function $z_{i+1} = z_i^2 + c$ can also be plotted for all z_0 and generates the Julia set of the parameter constant c. These Julia sets have a great variety of forms for different values of c:

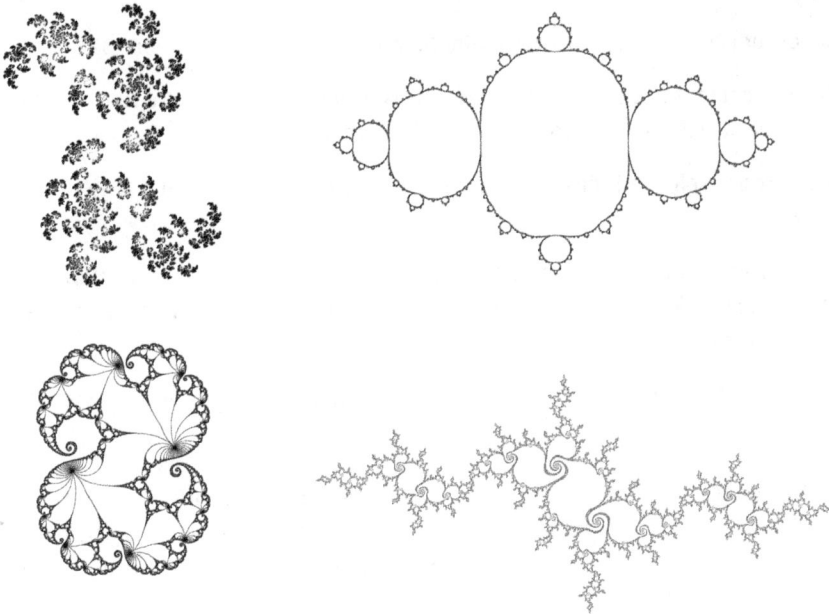

Figure 3. Julia sets for four different values of c. Top-left is disconnected, so c is outside the Mandelbrot set, the others are connected.

When the sets are fully connected, c is inside the Mandelbrot set, conversely when the sets are disconnected "dust," the location of c is outside the Mandelbrot set. The critical values between these two cases seem to occur precisely along the border of the Mandelbrot set. As the Mandelbrot set locally approximates the shape of each Julia set along its border, it inherits this variety of Julia set forms all within a single shape.

This stunning diversity of visual forms is largely responsible for bringing the topic of scale symmetry to the public's attention. It has sometimes been called the most complex shape in mathematics or the world's most difficult maze! These are not serious statements, but the complexity of this set in comparison to the simplicity of its definition is certainly striking. It also has many interesting properties that make it a unique and fascinating structure both mathematically and visually. This makes it a rare example of an object that appeals to mathematicians, artists and hobbyists alike.

The Mandelbrot Set in Six Facts

1. It is a mimic. Each small region along its border mimics the shape of its corresponding Julia set, giving it a rich variety of forms.

2. Its border is densely packed with mini Mandelbrots (dubbed minibrots).

3. It is connected, so each of these minibrots is connected by a tendril to the rest of the set in a tree structure, like apples hanging from their stalks.

4. The border itself has Hausdorff dimension 2, however the border's area is currently unknown.

5. It is universal: with rare exceptions, any algebraic maps of iterated functions include small Mandelbrot sets (or higher order equivalents) in their parameter space. See [16] for more information.

6. It is exactly the set of points where its corresponding Julia sets are connected.

But this chapter is not about *the* Mandelbrot set, it is about general Mandelbrot sets of the iteration $x_{i+1} = f(x_i) + x_0$ on some vector $x \in \mathbb{R}^n$ and for some iteration function f. The interest in these sets is in finding structures that inherit the variety and beauty of the Mandelbrot set.

There has naturally been a lot of interest in whether there could be a 3D analogy of the Mandelbrot set. On the surface, this doesn't seem too difficult. The $+z_0$ operation is just a translation, which can act in 3D as readily as it does on the complex plane. The complex squaring operation squares the magnitude of the complex number, and this also can be performed on a 3D vector. So far so good.

But the complex square also doubles its angle from the real line. This double-covers each concentric circle in the complex plane over itself, just like if you double up a rubber band and stretch it to its original circumference. The 3D equivalent of a circle is a sphere, and a point on a sphere has two angle coordinates, so one might consider simply doubling both coordinates. This produces what are called Mandelbulbs, which were first examined by Daniel White and Paul Nylander in 2009 [17]. There are several types of these because there is more than one set of spherical coordinates.

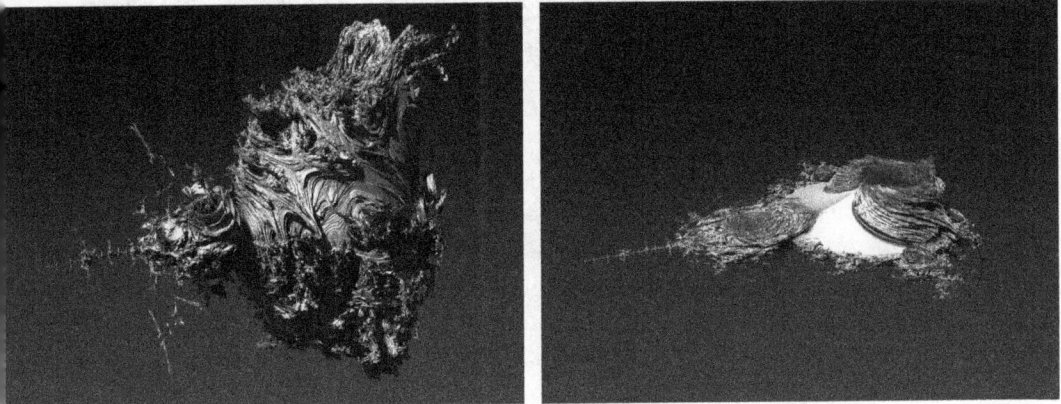

Figure 4.   Mandelbulbs using two different choices of spherical coordinate system.

Despite being interesting shapes in their own right, these Mandelbulbs don't quite have the beautiful intricacy one might expect from a three-dimensional Mandelbrot set, they contain large stretched and extruded areas. The problem is that doubling the spherical coordinates does not evenly scale each patch of the sphere, which is what we should expect if we apply the rubber band analogy to a 3D rubber sphere. We would like the rubber sphere to fold over itself and stretch evenly in all directions, so the transformation is conformal and small details are neither squashed or elongated by applying it. Unfortunately, such a transformation does not exist in 3D.

However, we can get some of the way there. For instance, we can quadruple-cover each sphere so that it is conformal over the sphere's surface. In other words, all small patches of the surface will scale evenly, or isotropically, in each tangent direction, but not equally in the radial direction. There are several folding patterns that achieve this but the most regular of these wraps a *spherical tetrahedron* over itself.

Figure 5.   The spherical tetrahedron wrap shown for one face, which is split into four colours (left). It grows (middle) until it occupies the full sphere as the four new faces of a spherical tetrahedron (right). This transformation is applied to each face with the edges remaining connected.

In order to wrap this spherical tetrahedron conformally, we make use of two facts: firstly, all functions on a sphere can be mapped conformally to the complex plane by a *stereographic projection* and back again by its inverse. Secondly, on the complex plane, any analytic function on complex numbers is conformal. In particular, the quotient of two complex polynomials is conformal (apart from at its poles).

The quotient in question is:

$$g(z) = \frac{-z}{2\sqrt{2}} \prod_{j=0}^{2} \frac{z - \sqrt{2}e^{2ji\pi/3}}{z - \sqrt{1/2}e^{(2j+1)i\pi/3}}$$

It is fully defined by its four zeros at the projection of the four tetrahedron corners and its four poles at the projection of the four tetrahedron face centres. In the formula, the three numerators inside the product operator and the single numerator outside make up the zeros, the three denominators and an implicit one at infinity make up the poles.

By the way, since this function is itself conformal, we can render it on its own as a type of 2D Mandelbrot set:

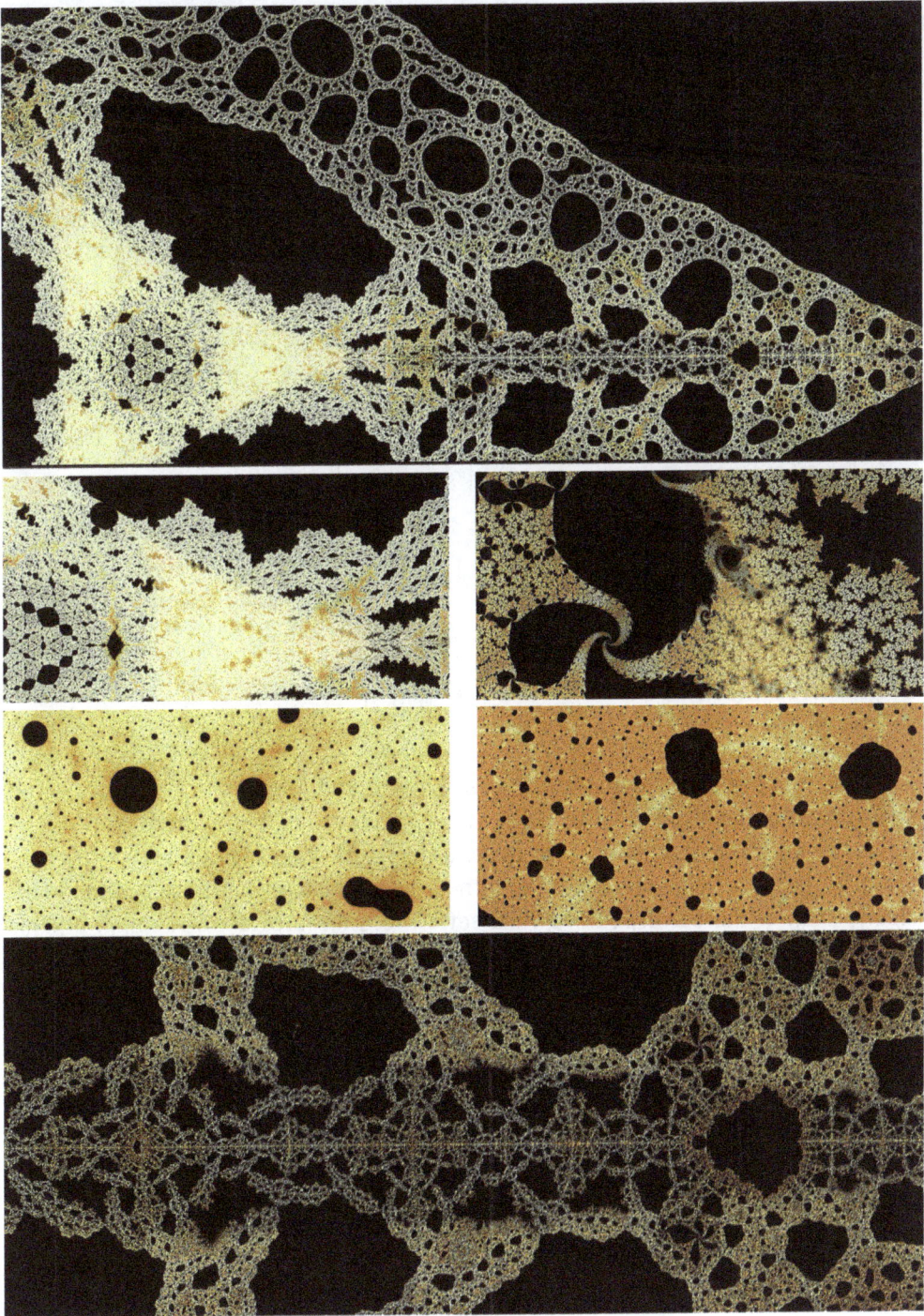

Figure 6.   Zooms of the tetrahedral Mandelbrot set $z_{i+1} = g(z_i) + z_0$. Top shows the main section.

There is a great deal of variety in the resulting shapes, and it looks quite unlike the standard Mandelbrot set. In particular, it is not a tree structure but sponge-like with many holes of various shapes. Some areas look like a cloud of asteroids, others more like chains and others like leaves. It is quite an enchanting exercise zooming in and finding new areas in this set.

Anyway, we want to use this function in 3D. We can do so by projecting each concentric sphere onto the complex plane, applying $g(z)$ and projecting back onto the sphere before applying the simpler 3D operations, which square the radius and add x_0.

3D Construction

For the iteration $x_{i+1} = f(x_i) + x_0$, the function $f(x_i)$ expands to:

1. Extract radius: $\qquad\qquad\qquad\quad r = |x_i|$

2. Project onto complex plane: $\qquad z_i = \frac{x_x + i x_y}{r - x_z}$

3. Apply tetrahedral transform: $\qquad z_{i+1} = g(z_i)$

4. Project to *Riemann sphere*: $\qquad x_{i+1} = \frac{\left(2 z_R, 2 z_I, z_R^2 + z_I^2 - 1\right)}{z_R^2 + z_I^2 + 1}$

5. Square the radius: $\qquad\qquad\quad x_{i+1} = r^2 x_{i+1}$

The resulting shape has a trilateral symmetry and appears to be a connected structure. Like the Mandelbulbs, it has some flatter areas, but without as much of the extruded, stretched regions. There are a variety of shapes visible such as branching trees, but it can also be viewed from the inside, which reveals spirals, often more clearly seen in the corresponding Julia sets. These features are sweeping, organic and attractive in appearance, and the trees and spirals are reminiscent of the 2D Mandelbrot set, offering a glimpse of what these Mandelbrot set features might look like in 3D.

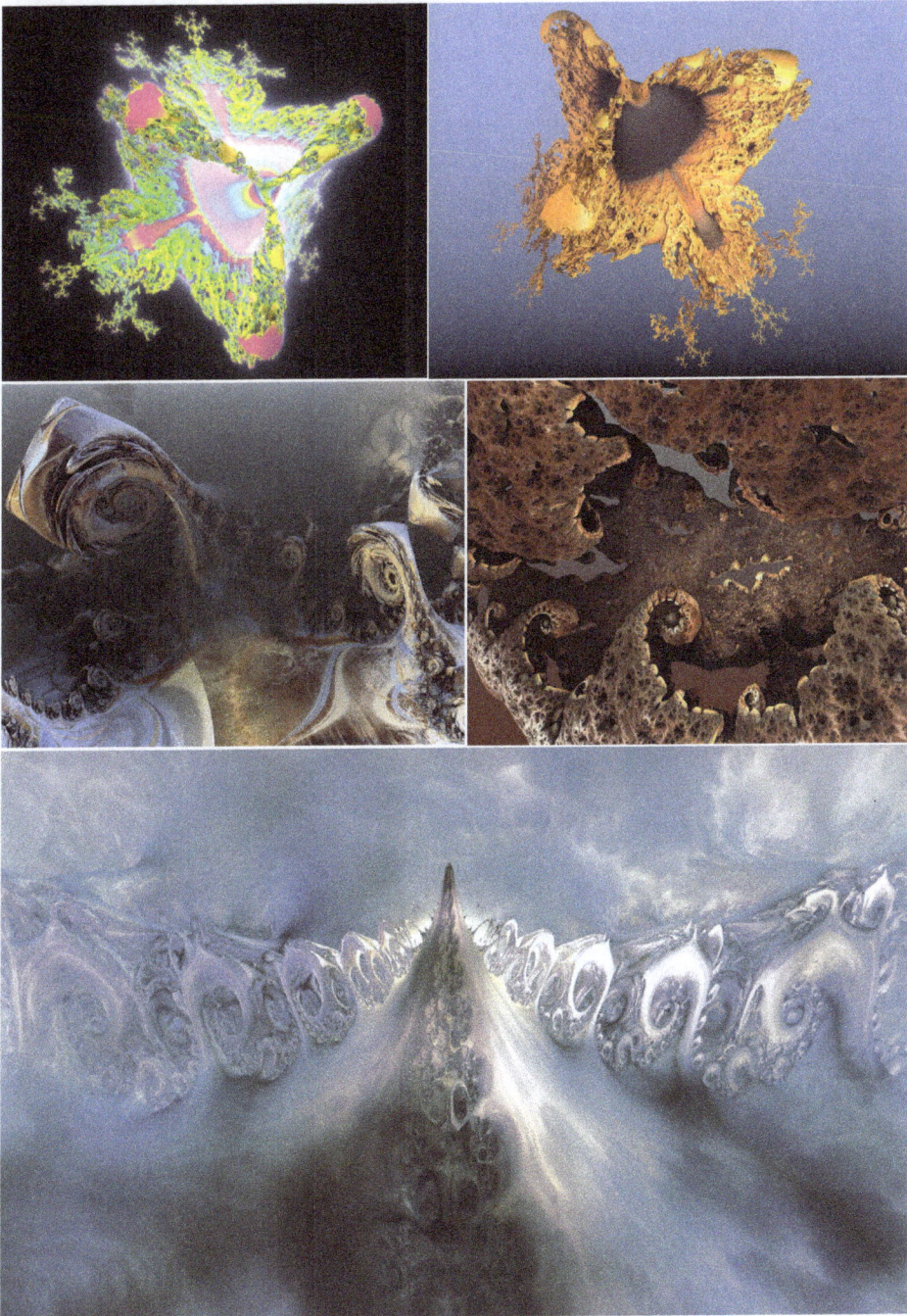

Figure 7.   Top: the full set. Bottom: inside render of spirals. Middle: spirals in the Julia sets.

There is still much to explore with this sort of 3D extension of the Mandelbrot set, for instance, the transformation is not conformal in all three directions only laterally. The same folding pattern could be made more conformal by finding a function that maximises how isotropic the scaling is in all three axes. Such a function is unlikely to be analytic, but we should expect a more conformal transformation to provide even more rich details in the resulting shape. The optimal function may not be smooth however, so the resulting shape may have sharper features.

The reason there is no conformal transformation in 3D that multi-covers the space is because, according to *Liouville's theorem*, there are only four possible conformal transformations in 3D. They are translation, rotation, dilation, and a sphere inversion combined with a reflection. Combinations of these four are also conformal — these are the *Mobius transformations* — but no combination of these will multi-cover the space as they are all single-cover, one-to-one functions.

However, if we relax the conformal requirement and allow small shapes to be reflected, then these are the *shape-preserving* transformations defined in Chapter 4, and the set of such transformations now includes reflections around planes.

When just one side of the plane is reflected, we call this a 3D fold. So, we can now translate, rotate, scale, invert *and fold* around arbitrary planes as desired, and the result will be *shape-preserving* everywhere apart from exactly on the fold surfaces. This transformation is the most general of the different transformations that we call scale-symmetric.

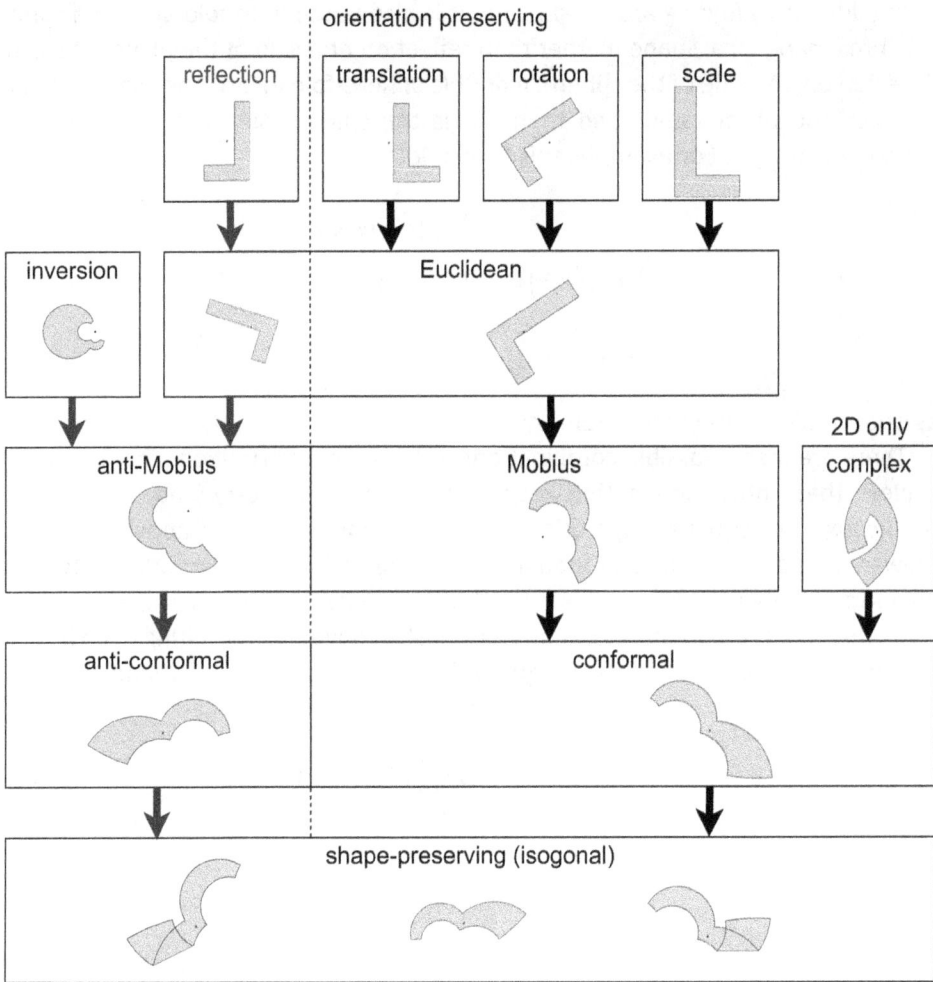

Figure 8. Increasingly general scale-symmetric transformations. Here, "complex" refers to any complex polynomial on the complex plane.

The 3D folding operation *double-covers* the space, which can then be scaled up. To generate a Mandelbrot set with this approach, the idea is that points close to the origin should be repeatedly folded inwards, but distant points escape towards infinity with each iteration. This qualifies it as an *escape time* fractal like the Mandelbrot set. A simple such folding is the *box fold*, which acts on each axis *i* like so:

$$f_{box}(x)_i = \begin{cases} 2 - x_i : x_i > 1 \\ -2 - x_i : x_i < -1 \\ x_i : \text{otherwise} \end{cases}$$

In addition to folding around a plane, it is also possible to fold around a sphere while preserving local shape. Rather than reflect on one side of the plane, the space is inverted on one side of the sphere. A double sphere-fold inverts the space between the inner and outer radius and then scales the space inside the inner radius to maintain continuity. For example, this "ball fold":

$$f_{ball}(x,r) = x \cdot \begin{cases} 1/|r|^2 : |x| < r \\ 1/|x|^2 : |x| < 1 \\ 1 : \text{otherwise} \end{cases}$$

has outer radius 1 and inner radius r.

There are many possible combinations of these transformations, but one of the simplest that shows interesting results has cubic symmetry and is called the *Mandelbox*. The squaring operation for the Mandelbox is the above "box fold" followed by the "ball fold" and then a scaling. The ball fold's inner radius is typically 0.5, and the scale parameter s is set to −1.5 in the examples shown here.

By the way, the sign of this parameter is just a convention resulting from how we perform the box fold, so we could equally define it as the scale 1.5 Mandelbox if we wished.

$$\text{Mandelbox: } x_{i+1} = s\, f_{ball}\left(f_{box}(x_i), 0.5\right) + x_0$$

$s = -1.5$:

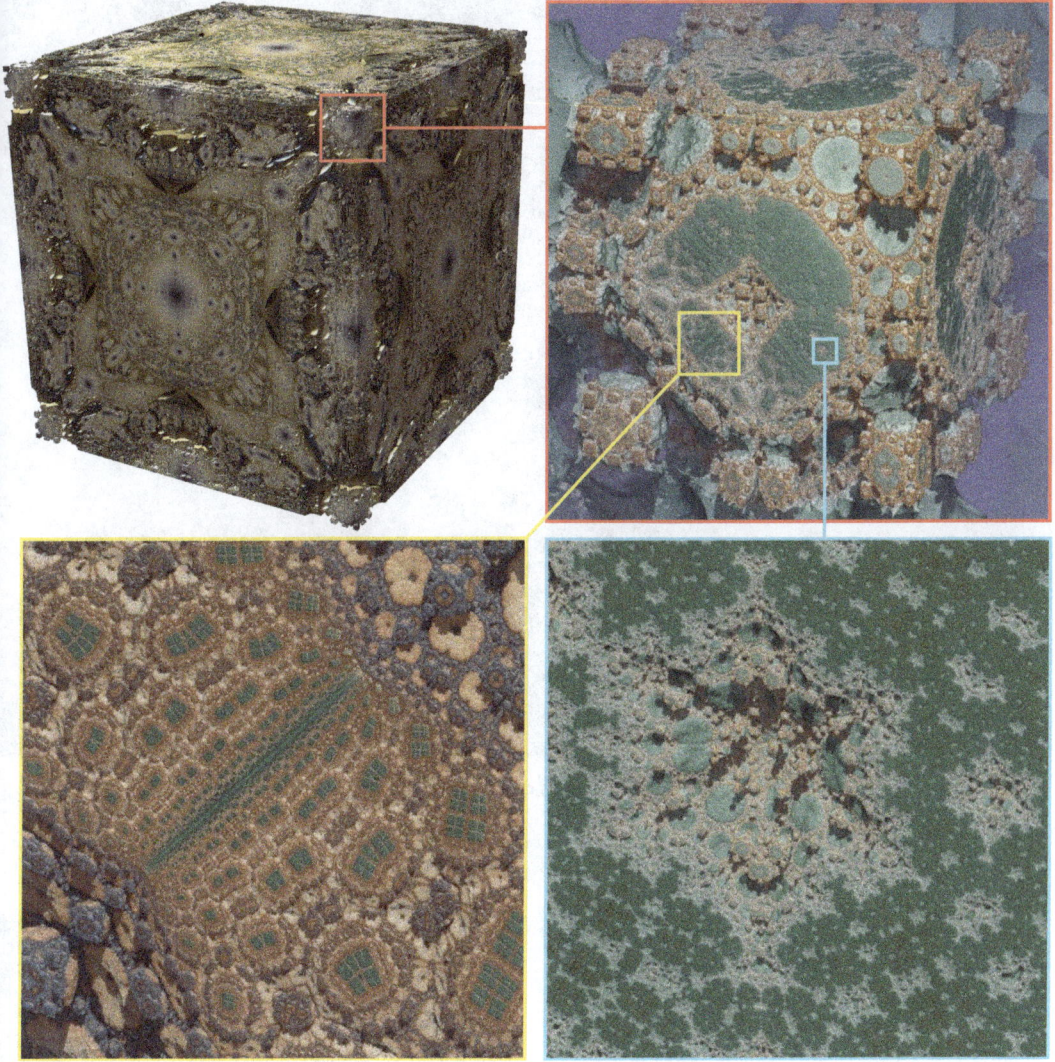

Figure 9.   Top-left: the Mandelbox. Right: zoom of corner reveals a similar looking "mini-box" with an *Apollonian gasket* on its edges. Bottom-left: pattern appears to tend to a 1D *Cantor set* at the line. Right: an area similar to a *Koch snowflake*.

Figure 10.   Deeper inside the Mandelbox is more complex rather like an alien space station. The "solar panels" mimic 2D *Cantor sets*.

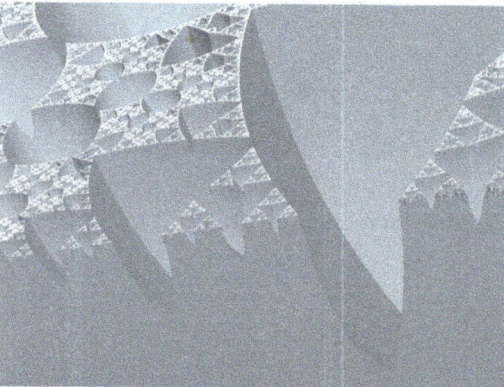

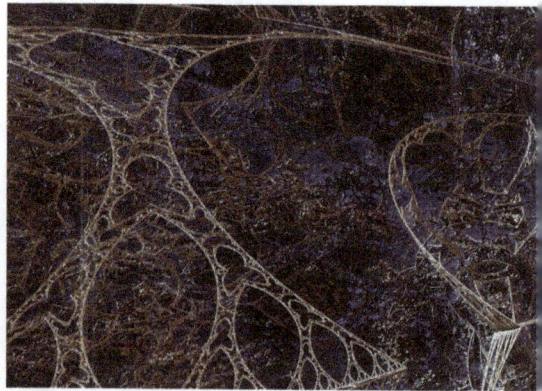

Figure 11.   Left: more mimicry, a *Sierpinski triangle*. Right: tree-like wire structures.

Figure 12.   Deeper within the Mandelbox, a tree structure built from scrap metal.

Figure 13.   Bush-like shapes growing from a girder-like structure. Combining man-made and organic elements.

$s = 2$:

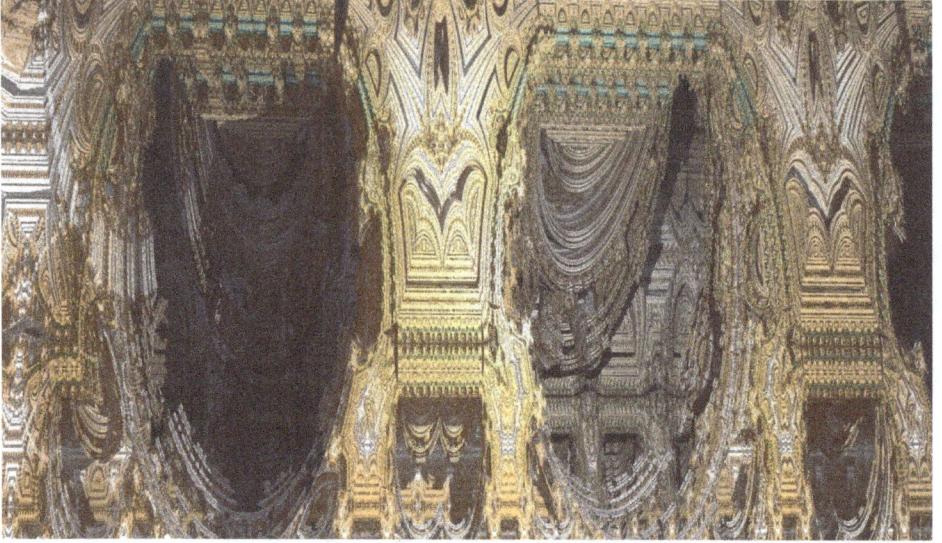

Figure 14.   The positive scale Mandelbox ($s = 2$ here) is more architectural in its features, showing a pillared entrance.

Figure 15.   The $s = 2$ Mandelbox from inside has the suggestion of an ornately decorated theatre.

With this type of Mandelbrot set, we have sacrificed smoothness to allow transformations that are shape-preserving. The results are completely different to the previous example. The folds produce a more architectural style, and the shape preservation means there is a lack of stretched and extruded areas, and so detail exists everywhere. The colouring in the above images is artistically chosen, but it is a function of the number of iterations required to find the surface, so the colouring has a geometric consistency.

The Mandelbox has many of the properties of the 2D Mandelbrot set. It is dense in its centre [18], it contains mini-boxes in the corners, just like the Mandelbrot's minibrots. It has a great deal of variety in its shape and its Julia sets are *shape-preserving*. An interesting thing to discover about this fractal is that it appears to contain approximations of many other well-known fractal shapes embedded within it. There are areas similar to the *Sierpinski triangle*, *Menger sponge*, *Apollonian gasket*, *Cantor set*, *Levy C curve*, fractal trees and 2D *Cantor dust*. I am sure there are many others too. This mimicry is certainly an interesting topic to delve into further. For instance, we might ask what range of fractals it approximates, and whether there is a formula that mimics more of them.

Despite the beautiful geometry that it displays, the Mandelbox is still lacking many of the mathematical properties of the Mandelbrot set as listed at the start of this chapter, so we cannot call it a close 3D analogy. Even from a visual perspective, it lacks a subtle but important feature. In the 2D Mandelbrot set boundary you can always find a nearby miniature Mandelbrot (minibrot) and use this to zoom into a completely different feature of the set. With the Mandelbox this doesn't appear to happen. If you zoom into the girder-like area on the edge of the Mandelbox, you cannot find mini-boxes, likewise if you zoom into one of the mini-boxes in the corner of the Mandelbox then you cannot find a girder-like area along its edge to zoom into. In this sense the variety in the Mandelbox is more strictly separated by location.

Also, despite the Julia sets being *shape-preserving*, the Mandelbox itself is in fact not. So, small shapes can become elongated. This is a consequence of the $+x_0$ part of the formula, it allows a small shape to be added to a reflected copy, which causes the elongation.

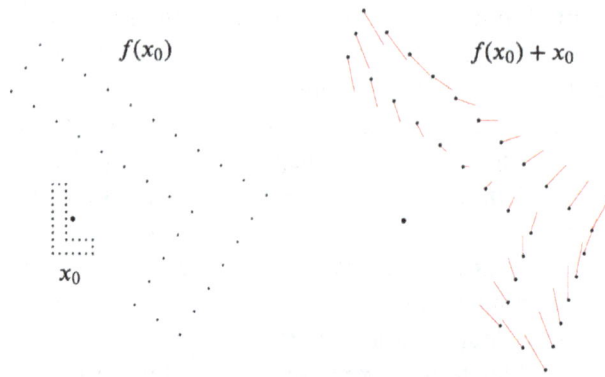

Figure 16. For values of x_0 in an L shape, the reflecting transformation $f(x_0)$ is shape-preserving, but $f(x_0) + x_0$ elongates.

Of course, those people hoping for a "real" higher dimensional Mandelbrot set are out of luck. Not only are conformal multi-covers of space impossible in 3D, they are impossible in all higher dimensions as well as a result of *Liouville's theorem*. In addition, the associative division algebra that characterises complex numbers only exists for the 1D reals, 2D complex numbers and the 4D quaternions according to the *Frobenius theorem*. While quaternion Mandelbrot sets are possible, they also have an extruded appearance because they are still subject to *Liouville's theorem*, so are not conformal. However, the lack of a direct 3D analogy should not stop the budding explorer! Many of its interesting properties can be extended to 3D, just not all of them. So, in that sense, the Mandelbrot set remains more of an inspiration for finding interesting 3D fractals than a formula to replicate directly in 3D.

On these lines, there is more we can do. We noted that including reflections in the transformation is problematic as the function's Mandelbrot set is no longer *shape-preserving*. If we allow discontinuities in $f(x)$, then we can double-cover 3D space without reflections by cutting it at a plane and placing the two half-volumes over each other. Without reflections, the resulting Mandelbrot set remains conformal. Unfortunately, discontinuous transformations tend to produce sharp and discontinuous resulting shapes.

But rather than cutting the space each iteration, we could take two copies of the space and transform both by some transformations m_0 and m_1. Each x_0 therefore becomes two separate x_1 locations, and four x_2 locations and so on, branching each iteration like a binary tree. We throw away branches that exceed a chosen escape radius and then record the number of branches that remain after n iterations. This number can be visualised with a heatmap colouration for each x_0. This approach avoids discontinuities and it avoids reflections, so any conformal m has a conformal Mandelbrot set, and stretched-out regions are avoided. However, the word *set* is a misnomer here, as the result is an integer value per point, so it is a type of Mandelbrot *multiset.*

We will make m_0 and m_1 Mobius transformations as they are conformal in every number of dimensions. So, let's call this family the *Mobius multisets,* and start by looking at the 2D case. There is one such *Mobius multiset* for each possible pair of Mobius transformations.

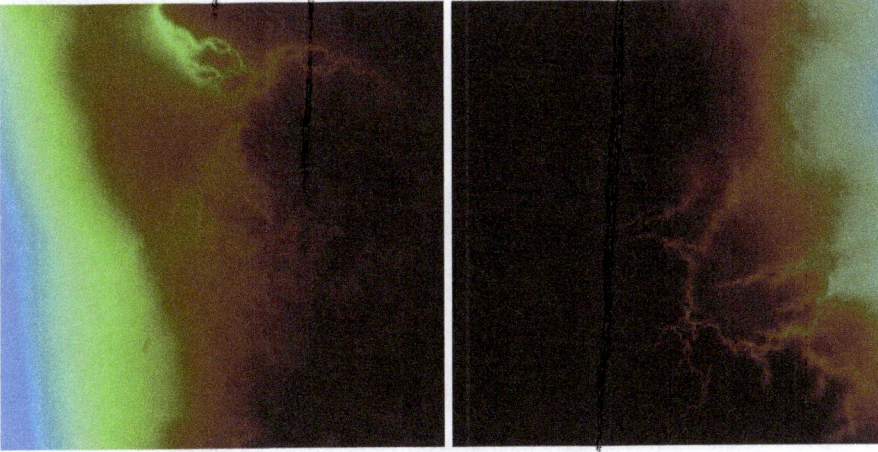

Figure 17.   Mobius multisets: looking very stormy.

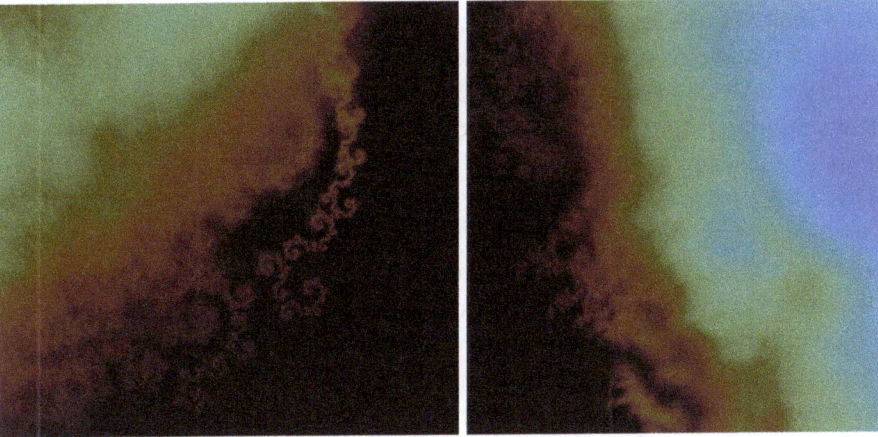

Figure 18.   Left: geometric shapes reveal themselves. Right: organic looking cloud cover.

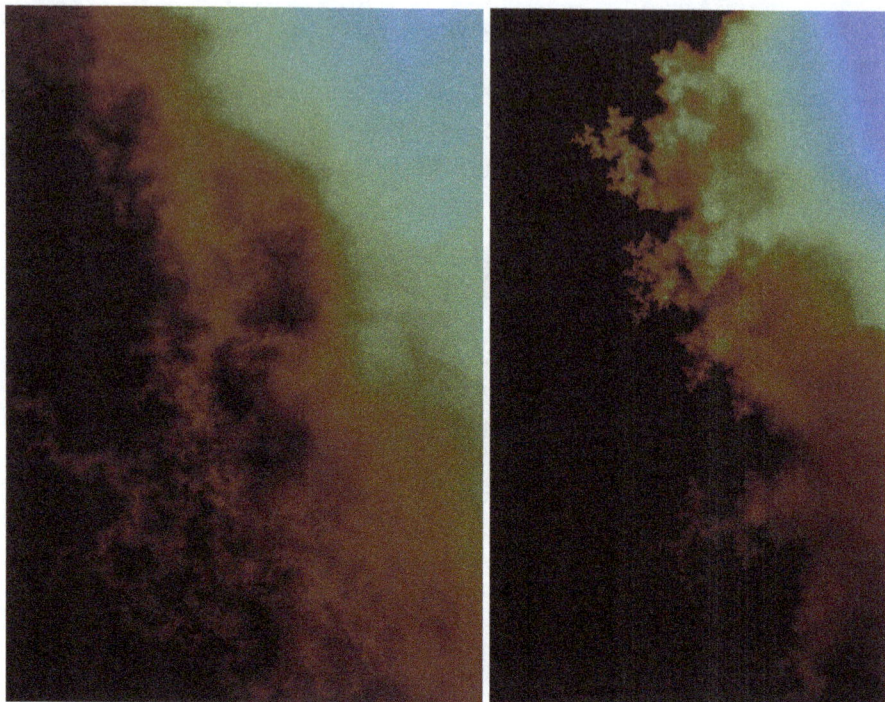

Figure 19.   Above-right: the emergence of 3D-looking shapes with soft lighting. How this happens is a mystery.

As a multiset, these maps have a very different appearance to traditional fractals, with a moody, or dramatic storm-like character. Some areas appear like lightning, others like patchy or swirling cloud, in other areas, geometric spirals and curlicues emerge from clouds. This seems far removed from the hard and artificial features of Euclidean geometry.

One of the enticing things about exploring new ideas such as these is that you can never quite be sure what mystery will turn up next. For example, the figure above-right (Figure 19) has an unexpectedly three-dimensional appearance. What's more, it appears to be rendered with occlusion and quite a realistic looking illumination model. How this could emerge from a heatmap colouring of a multiset is a bit mysterious. It would be a difficult but interesting mathematical exercise to discover how simple branching Mobius transforms create this illusion of 3D geometry.

When the Mobius transformations have no translation component, the multiset appears to be connected. In the following case, the image also has some similarity to the Mandelbrot set, with curly spirals and branching trees.

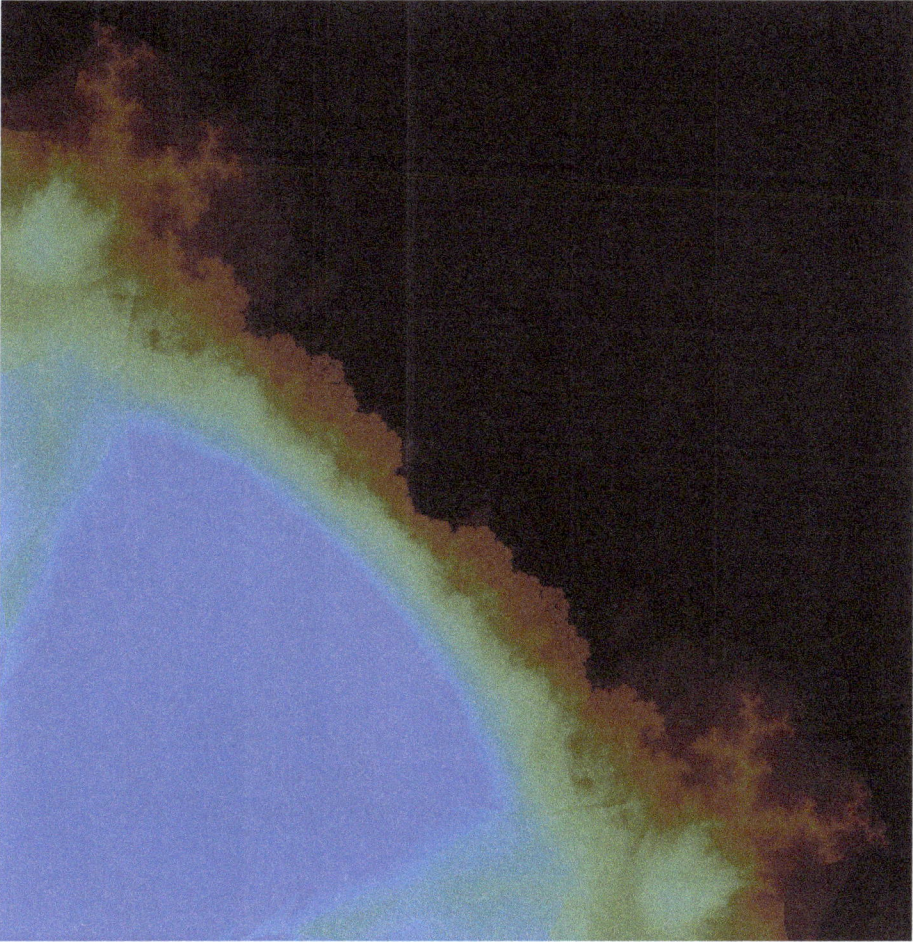

Figure 20. Zero translation symmetric Mobius transform shows a progression in the shapes from centre to corner.

Mobius Multiset Algorithm

For each point x_0, we grow a set from $X = \{x_0\}$ according to:

$$X_{l+1} = \bigcup_{x \in X_l} e(m_0(x)) \cup e(m_1(x))$$

with the escape criterion for escape radius R:

$$e(x) = \begin{cases} \{x\} & \text{if } |x| < R \\ \{\} & \text{otherwise} \end{cases}$$

and Mobius transform:

$$m_j(x) = s_j \text{ref}\left(|b_j|^2 \frac{x - b_j}{|x - b_j|^2} + b_j \right) + d_j + x_0$$

using the reflection:

$$\text{ref}(x) = x - 2r_j(x \cdot r_j)$$

The resulting multiset is the set of all points x_0 together with their corresponding multiplicity $|X_n|$, which is the size of the set after n iterations.

Programmatically, x_0 are the pixel locations and the integer $|X_n|$ is mapped to the pixel colour.

In the above image, we use:

$$n = 16, R = 4, b_0 = (1,0), b_1 = (0,1), r_0 = (-1,1)/\sqrt{2}, r_1 = (1,-1)/\sqrt{2}$$

$$s_0 = 2, s_1 = 2, d_0 = (0,0), d_1 = (0,0)$$

These multisets are reminiscent of the *Buddhabrot set* [19] and the *Littlewood fractal* [20]. The first is generated by plotting every z_i of the Mandelbrot iterations, and the second is a plot of the roots of Littlewood polynomials, which are those with coefficients of 1 or −1.

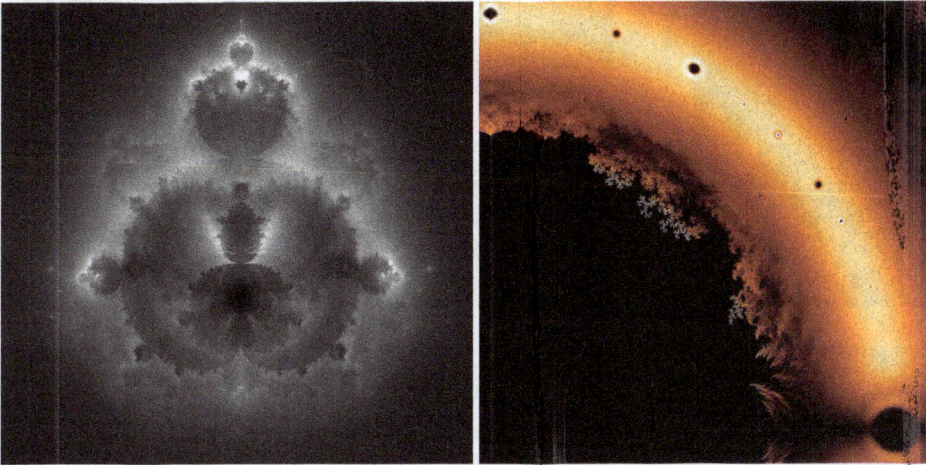

Figure 21. Buddhabrot (left) and Littlewood fractal (right) are also multisets with a softer appearance.

In both of these, the fractal patterns also vary with location and they can display many different familiar fractals within one map. But unlike the Buddhabrot and Littlewood fractals, *Mobius multisets* can be easily applied in 3D. Simply use three 3D Mobius transformations at each branch point instead of two 2D ones, an example of this is shown at the end of the chapter.

While the pictured multisets have a cloudy and sometimes blurry appearance, this is a consequence of the limited number of iterations. As the iterations increases, the results tend towards a precise structure. But with increased precision, we have to be extra careful with our choice of escape criterion $e(x)$.

Unlike the Mandelbrot set, the Mobius iterations don't escape to infinity for outside points, and they don't converge to any single location either. Instead, they converge to a region, so the ideal choice of escape criterion is based on proximity to this region. The simplest general way to find this region is to iterate a random set of points within the view window, using a random choice between the two Mobius transformations. These points will tend towards the escape region with increasing iterations.

For the Figure 20 multiset, the outside points tend towards a region in the top-right corner, so it is sufficient to use the distance in this direction as the selection criterion. Doing this, and doubling the iteration count to 32, reveals the underlying, precise structure. Here, we show a rotated close-up of Figure 20's bottom-right quadrant:

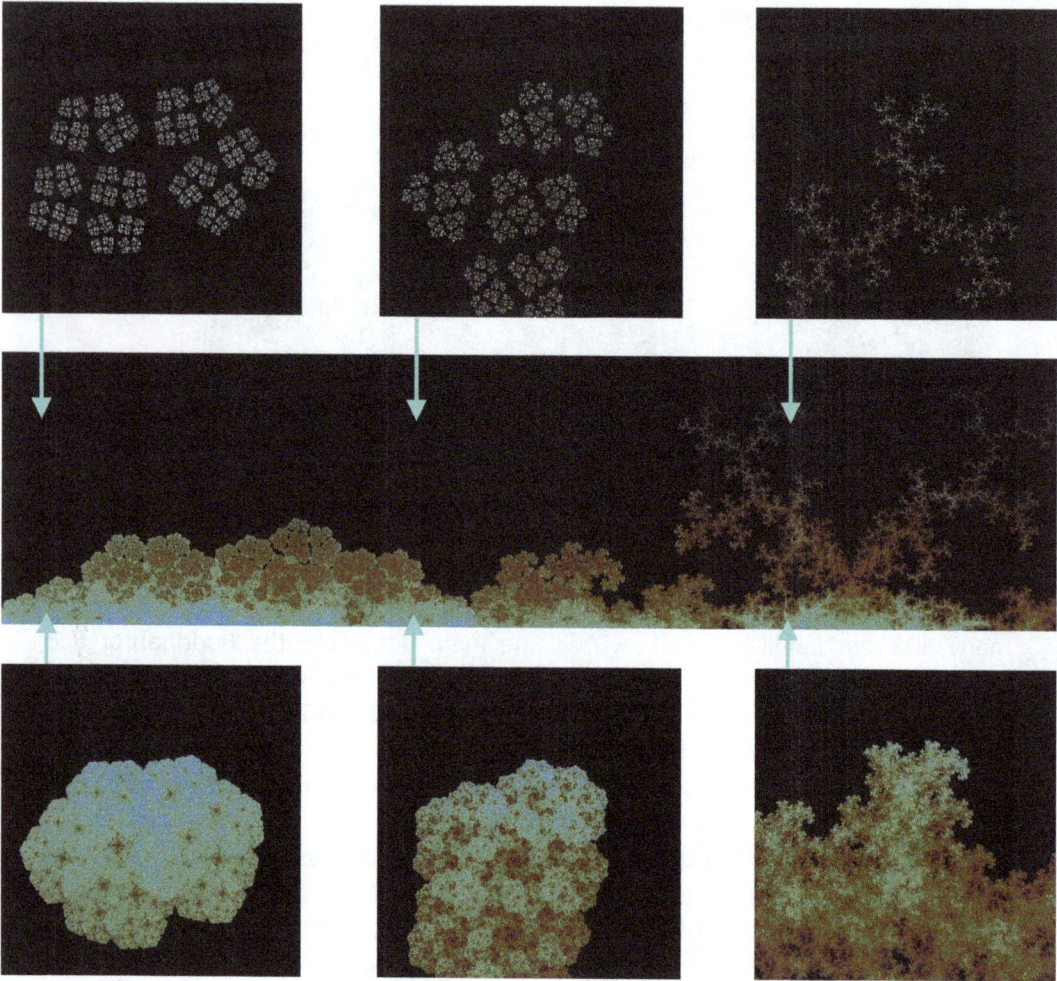

Figure 22. Zoom of part of Figure 20. Using 32 iterations removes the cloudy areas. Surrounding boxes show Julia multisets located at the arrow heads.

The surrounding boxes show Julia multisets, where we replace x_0 in $m_j(x)$ with c (the location of the cyan arrow heads). As with the Mandelbrot set, the Julia multisets outside of the Mobius multiset appear to be disconnected "dust," those inside (bottom row) seem to be fully connected. The extra detail helps clarify the variation in the structure along its length, from blocky shapes on the left to spirals and then to dendritic patterns on the right.

If desired, we can remove the multiplicity, and so collapse the multiset into a single and precise limit set:

Figure 23.   Plotting each point with at least one path that remains bounded, a set is generated. Note the similarity in top-right to the tridendrite fractal of Chapter 2.

This family of limit sets is an interesting area to pursue, as these sets have more properties of the famous Mandelbrot set than any other in this chapter. They are everywhere conformal, varying with location, and they are precise limit sets — in this case, it may even be a connected set. What's more, they are extendible to any number of dimensions.

The previous example defined the two Mobius transformations through "bend vectors" (b_i) of (1,0) and (0,1). If we use three Mobius transformations with vectors (1,0,0), (0,1,0), and (0,0,1) instead, then we can generate a similar structure but in 3D.

Taking the average of these vectors as our vertical axis, the resulting shape shows a landscape with a detailed triangular structure on it.

Figure 24.   3D Mobius multiset, showing its large-scale triangular structure.

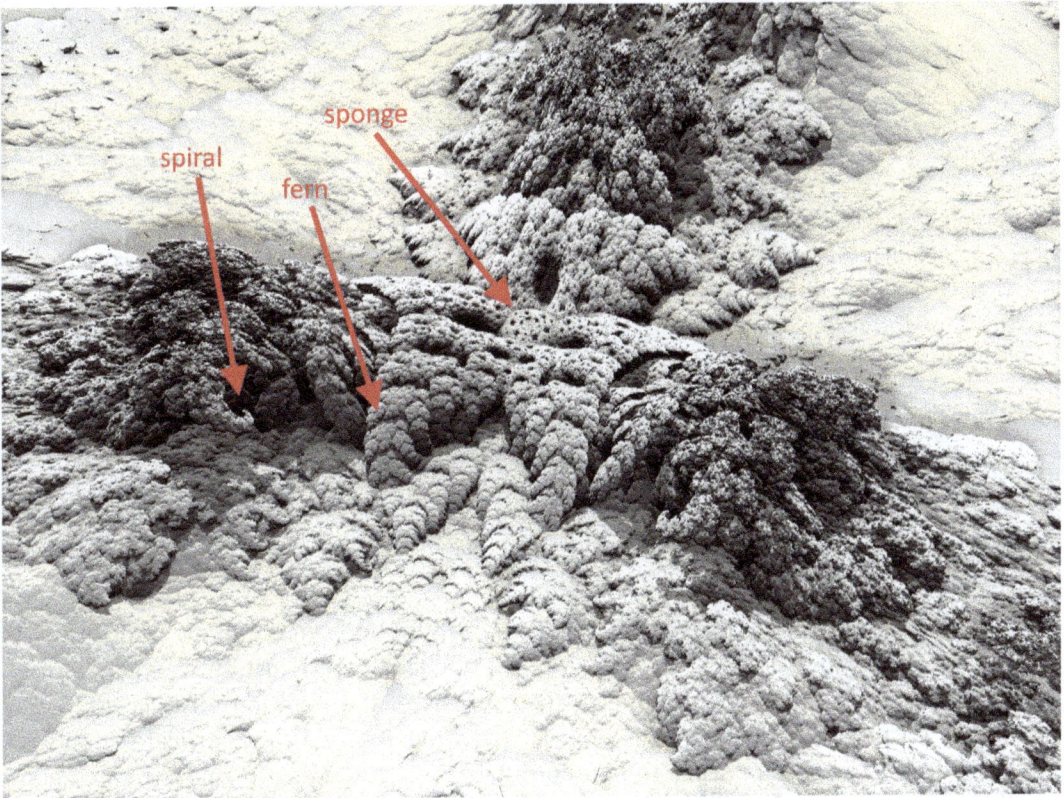

Figure 25.   3D Mobius multiset, close-up of centre of the triangular structure, with some noticeable structures labelled.

The structure is quite complex, featuring both flat and bumpy regions. The close-up of the centre in particular gives a nice example of 3D variation. The very centre is sponge-like and riddled with holes, just out from this are structures with a fern-like appearance, and a little further out, some spiralling structures can be seen. It certainly looks like an interesting place to explore further.

3D Mobius Multiset Algorithm

The method is the same as for the 2D case, but using three Mobius transformations rather than two. So, we grow the set X from $X_0 = \{x_0\}$ according to:

$$X_{i+1} = \bigcup_{x \in X_i} e(m_0(x)) \cup e(m_1(x)) \cup e(m_2(x))$$

with the selection function for threshold R:

$$e(x) = \begin{cases} \{x\} & \text{if } x \cdot V < R \\ \{\} & \text{otherwise} \end{cases}$$

and where m_i have the same formula as the 2D case. Any point with $X_n \neq \{\}$ has not escaped and so is considered part of the set. The images were generated by ray-stepping using this inside–outside test.

In the example, we use:

$$n = 27, R = 0.5, b_0 = (1,0,0), b_1 = (0,1,0), b_2 = (0,0,1), r_i = b_i - (1,1,1)/3$$

$$V = (1,1,1)/\sqrt{3}, s_i = 2, d_i = (0,0,0), \text{ for } i = 0,1,2$$

★

So, looking back on the examples in this chapter, one can definitely notice the greater variety in the structures than in previous chapters, not just between images but within each image. This is due to the $+x_0$ part of the iteration formula that is characteristic of Mandelbrot sets. This variation together with the intricate details tends to give a sense of scale and grandness to the structures. Moving the camera through them often feels like exploring some sort of abandoned alien complex or an undisturbed world. The viewer is always the passive observer of a static scene.

In the next chapter, we will change that entirely by introducing interactivity. We therefore begin to include the viewer within the mathematical universes that are created.

Chapter 6

Cellular Automata

The principles of scale-symmetric geometry can be applied to interactive and animating structures by generating them through a system of local rules rather than through iterated functions. Such rule-based systems are usually referred to as automata, a well-known example are cellular automata such as Conway's Game of Life. Cellular automata apply rules over a repeating lattice of cells, making them translation symmetric, but they are not scale-symmetric by design, even though occasionally certain rules can give scale-symmetric results. In this chapter, we investigate cellular automata with inherent scale symmetry.

Conway's Game of Life

In this game, square grid cells are initialised randomly to be either dead or alive, then the same rule is applied to each cell based on the number of live cells of its eight nearest neighbours:

- Only live cells with two or three live neighbours survive, being neither underpopulated or overpopulated.
- Dead cells with exactly three live neighbours become live, as if by reproduction.

This simple set of rules (ruleset) is rich enough to make the system *Turing-complete*. So, any calculation that a Turing machine such as a computer can make can also be built in the Game of Life from a particular starting state of live and dead cells.

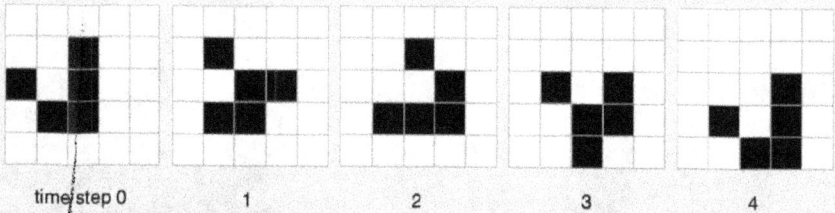

time step 0   1   2   3   4

Above is an example of repeated iteration of the Game of Life, creating a pattern of shapes that shifts diagonally, called a *glider*. Gliders are useful components for designing "machines" in the game, which process information.

The idea behind scale-symmetric automata is that the same ruleset is applied at multiple different scales, typically these scales are powers of two. So, if the ruleset applies on 1 pixel squares, then the same rule should apply on blocks of 2×2 pixel squares and also 4×4 pixel squares and so on. By extension, it should also apply on ½×½ pixel squares and all the smaller scales too, but usually there is little visual difference below the pixel level, so we will avoid this sub-pixel computation.

The first set of automata to look at do not animate like Conway's life, they produce static images by repeatedly doubling a low-resolution grid and applying rules to convert the low-resolution cells into the higher resolution pattern. We will call these *upscaling automata,* as they upscale the resolution, and they are automata as they apply a simple ruleset repeatedly. In this case, the system converts a very coarse starting seed, such as a random 4×4 image into higher resolution geometry. As with animating automata, they also allow a form of user interaction: at any point, the user can adjust this geometry at its current resolution and the resolution doubling will continue from this altered state. This allows for interesting user-guided designs, where an artist can guide the design and the automaton fills in the details and maintains its particular style.

The first such method considers the state of the four nearest neighbours on the parent (half-resolution) layer. We must then define a ruleset that converts the live or dead status of these four parent cells to a live or dead value for the child cell.

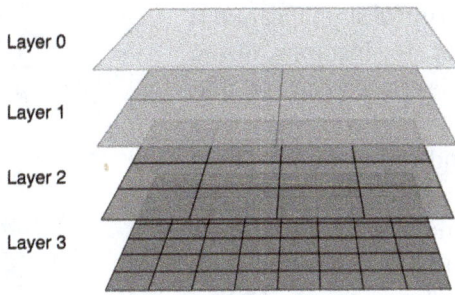

Figure 1.   Upscaling automata: each cell in layer *n* is set as a function of the nearest four cells in layer *n* − 1.

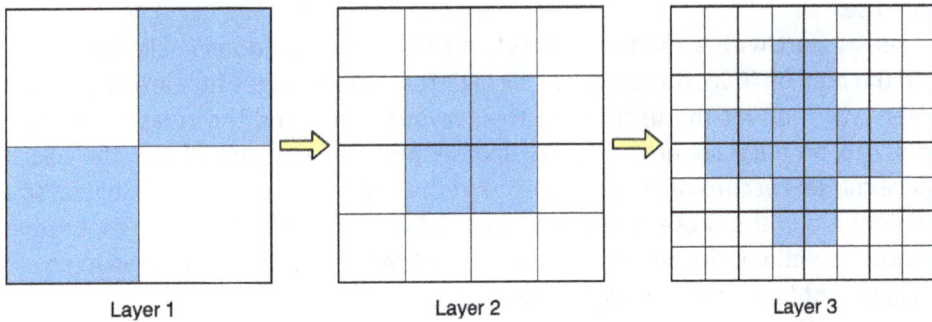

Layer 1	Layer 2	Layer 3

Figure 2.   Example ruleset starting with 2×2 seed: child cells set if at least two of the four nearest parent cells are set.

There are 2^4 permutations of the parent cell states, and each permutation determines one bit (live or dead) for the child, so each ruleset can be encoded in 2^4 bits. This means there are $2^{(2^4)}$ or 65536 different rulesets we could adopt, which is large number of rulesets to search to find something interesting. But more concerning is the growth rate, what if we looked at the nearest nine parents instead? Our search space would become $2^{(2^9)}$ which is over 10^{154} rulesets to look at, far too many to find geometrically interesting results. When the search space is this large, the results lack coherent patterns and are generally more random in appearance. For this reason, a large part of our search for interesting automata will be about finding ways to reduce the system rules to a smaller search space.

Probably the most fruitful simplification is to introduce symmetries, for instance, in the four-nearest-parents ruleset, we can see that the geometry of the child to the parents has a mirror symmetry around the diagonal. There are only 12 distinct parent states up to this symmetry, so there are now only 4096 rulesets to look at, rather than 65536. Even this number is large, so we still need an efficient means of searching the space. The method we use is a user interface that implements a select-and-mutate scheme to allow a user-guided search.

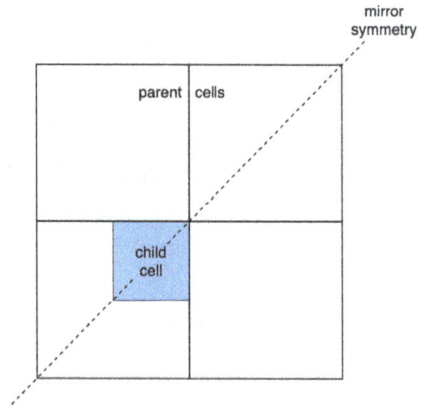

Let us start with a random 4×4 set of cells and six randomly selected rulesets from the pool of 4096, this produces six high-resolution images that are displayed on screen. We then let the user select their favourite set, and the rules are mutated slightly to produce six new rulesets and six new images. This allows the user to evolve rulesets according to the results that they find most interesting. Several small additions aid in the process: we can allow the user to restart the process, to revert previous selections, and we vary the mutation level across the six children to provide the option of both small and large changes.

Figure 3.   Select-and-mutate user interface. Top-left shows the randomly selected 4×4 seed, the seven remaining panels are randomly selected rulesets on the four nearest parent cells. When one is selected, it is moved to top-middle and a mutation of this ruleset is applied to the six lower panels, where the process can be repeated. Top-right shows the parent layers of the top-middle panel.

Here, we used a random 4×4 starting seed, but we can also specify initial starting shapes like a central square of 2×2 cells, or a horizontal line or a single cell. These all help inform us of how the automata act on different shapes.

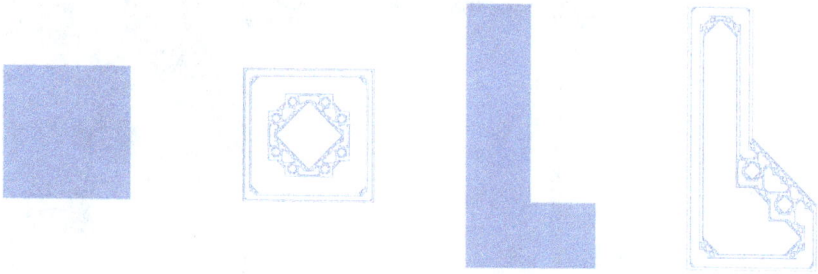

Figure 4.   The same ruleset as that selected in the prior image applied to two chosen seeds: a central square and an L shape. This time to a higher resolution by using more layers.

Applying the select-and-mutate method several times leads to some diverse patterns:

Figure 5.   Square-symmetric upscaling automata. The square symmetry is in the rules rather than the images. But we can see its effect within the structures, most clearly in the bottom-left image, but also in the top images. The bottom-right image contains what appear to be octagons, but they are not regular octagons, so they also have square symmetry.

A noticeable characteristic of all the images is the square symmetry within the structures, which is of course due to the symmetry of the subdivision rules. This means that shapes within the structure may also appear reflected around the horizontal or vertical axis, or rotated by 90°, but never in any other orientation, giving the structures an artificial appearance.

We should get more natural structures by allowing patterns to form in more than just four orientations. However, it is not immediately obvious how to extend these scale-symmetric automata to greater rotational symmetry such as hexagonal or octagonal symmetry because hexagonal grids do not line up cleanly with their larger versions and regular octagons don't tile at all. But it can be done in a slightly altered form.

A quasi-octagonal symmetry can be achieved with the square grid by rotating every other grid size by 45°. In this setup, each larger grid is scaled up by the square root of two rather than by two. The result is that the scale symmetry is now a scale and rotate 45° symmetry, so the two symmetries are combined. Under this geometric structure, each cell has two contacting parents at 45°. This is too few to produce interesting results, so the next largest symmetric neighbourhood uses the closest six parents to each cell. To reduce the search space, we constrain the rules to the rectangular symmetry of the structure, which allows for 2^{16} distinct rulesets.

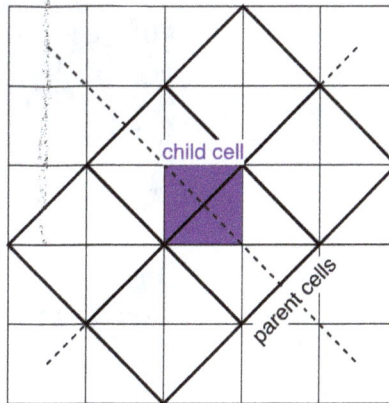

Figure 6.   Quasi-octagonal automata: the symmetry of the child cell to that of its six parents is rectangular symmetry (two reflections).

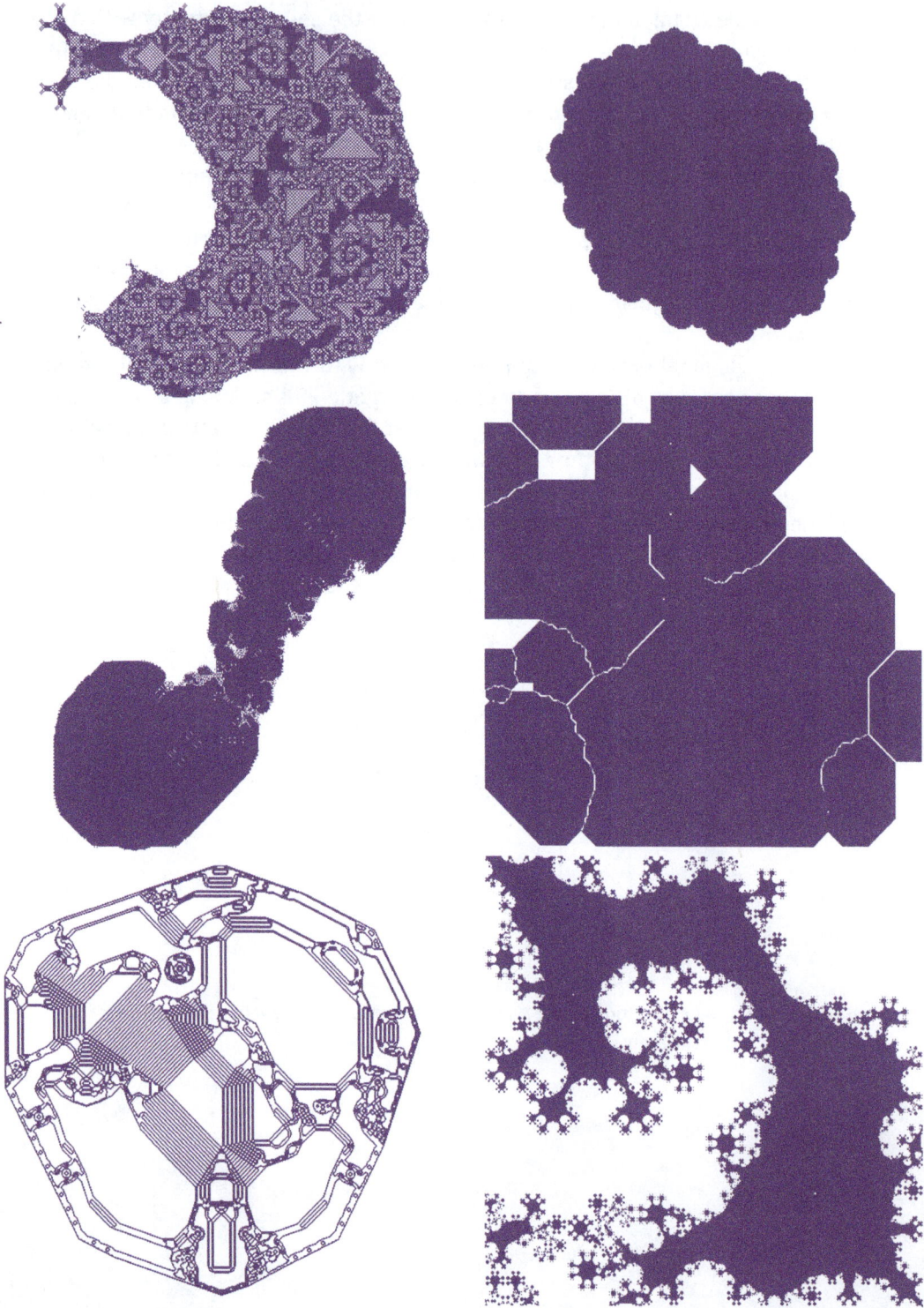

Figure 7.   A selection of the more interesting octagonally symmetric upscaling automata rulesets.

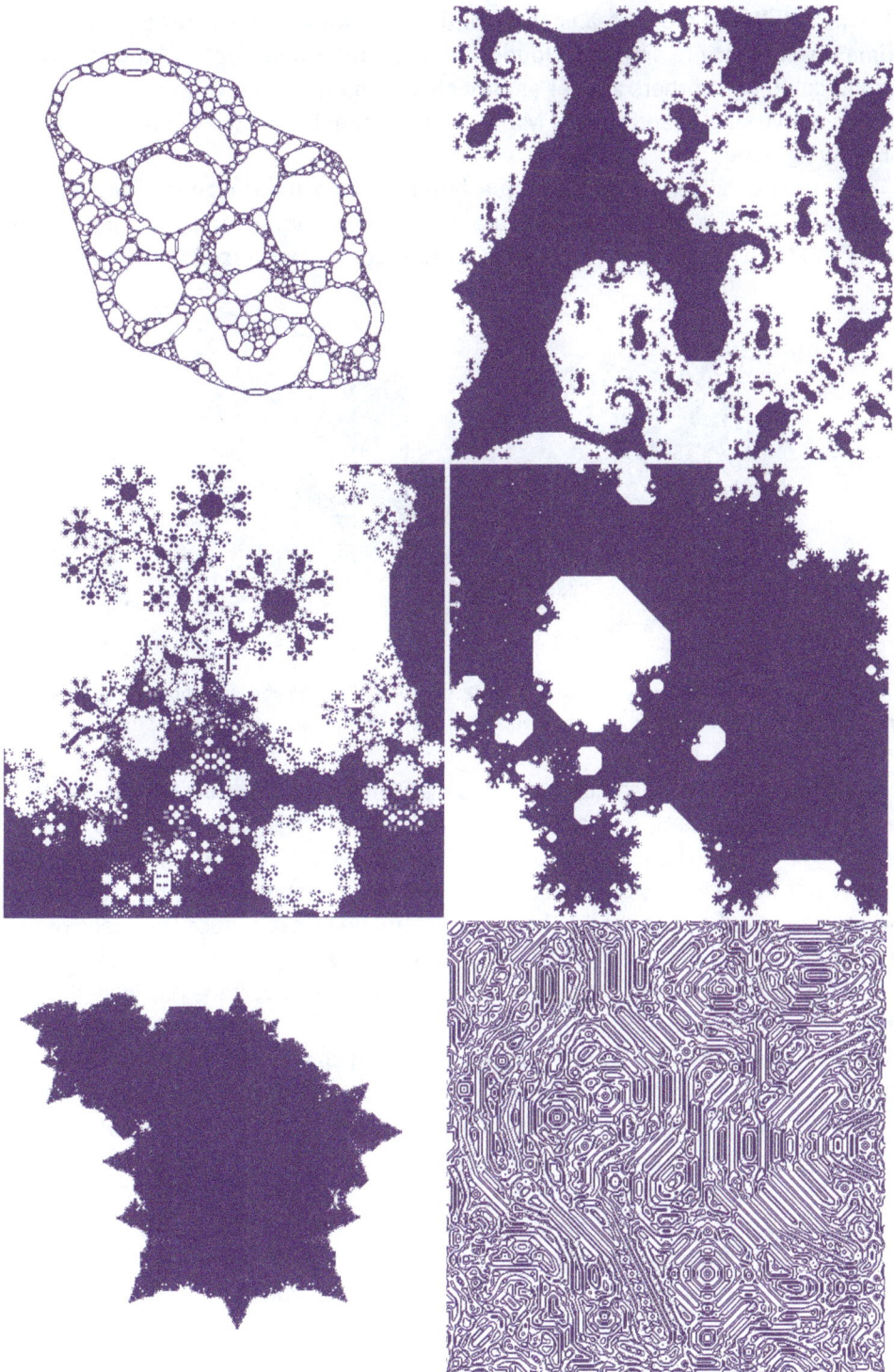

Figure 7. (*Continued*)

The results are more organic than for the rules with square symmetry, they show some resemblance to amoeba, clouds, bubbles, cracks and plants. Many are built of a solid colour, but others appear as though with a little shading, and some are patterned with triangles and spirals. Many of them seem to give the impression of living or growing things.

There is an alternate and non-quasi-method to achieve higher rotational symmetries in 2D. The method applies the axis aligned grid hierarchy in three dimensions and then renders only a diagonal cross-section of the results.

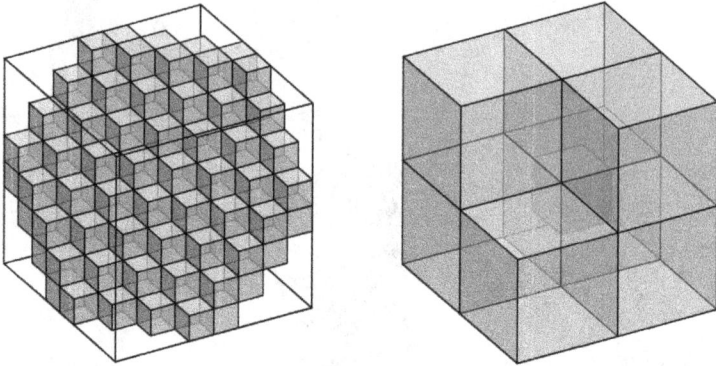

Figure 8.   Left: the cells in the diagonal cross-section of an 8×8×8 cubic grid show a hexagonal symmetry, these are the ones that are rendered. Right: the seven nearest parent cells of the green cell.

In this structure, each cubic cell has eight nearest parent cells, however these are not equally distant, so we ignore the farthest which only contacts the child cell at a single point. When we account for the symmetry of this arrangement of cells, we are left with 2^{40} possible rulesets. This is far larger than for the previous example and too large to hope to search the space fruitfully. We can reduce the space by adding an extra symmetry that we can call bit symmetry.

Bit symmetry means that the rules do not care if we invert the bits in all the cells, it will behave the same way. Translating to colours, it means that if you swap the colours in the initial seed pattern, then you will end up with the same high-resolution shape just with the colours swapped. This may seem like an unnecessary symmetry to impose on the automata, but in practice, it doesn't appear to overly constrain the results and it helps to drastically reduce the search space. In this case, we drop from 2^{40} to 2^{20} possible rulesets.

We reduce this to 2^{19} by requiring that the child cell is 0 whenever all seven parent cells are 0. This removes the unintended effect where every doubling of resolution the image is inverted, but otherwise has no other constraining effect.

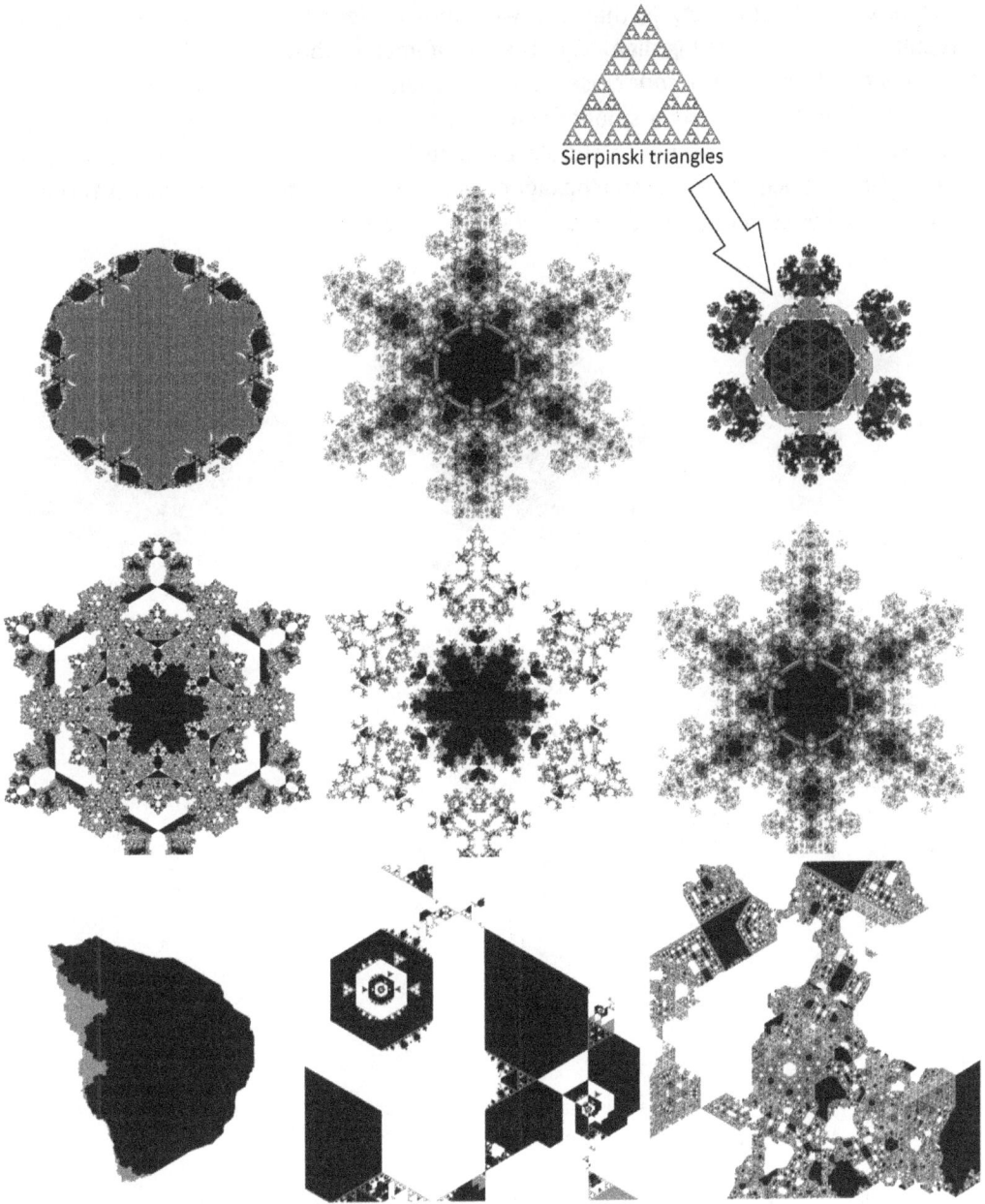

Figure 9.   Hexagonally symmetric upscaling scale-symmetric automata using a 4×4×4 seed with the 2×2×2 centre bits set. Bottom row: random 4×4×4 seed. Note that *Sierpinski triangles* are found in the top-right image and also in the bottom-middle image.

The results, as expected, show hexagonal symmetry, while retaining the same variety of forms that were seen with the quasi-octagonal method. A consequence of the hexagonal symmetry is that the well-known *Sierpinski triangle* shows up in results occasionally, as highlighted in the set of images above.

If we only look at the *four* closest parents, then we can remove the bit-symmetry constraint and it remains a smaller search space. The results do appear to have a simpler design to them, which should be expected of a smaller search space, and one even produces a very simple hexagonal snowflake structure that looks as though it could be constructed by a simpler *substitution rule*.

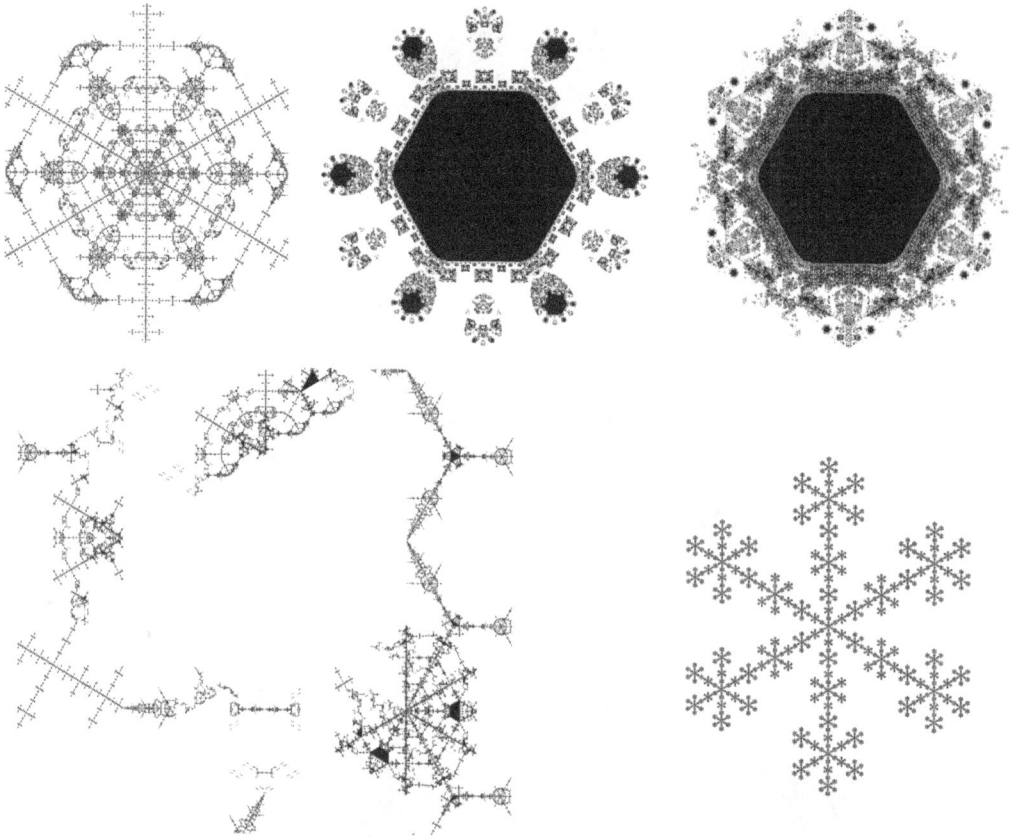

Figure 10.   Hexagonal symmetry automata using only the four closest parents. Bottom-left uses randomly set 4×4×4 seed, others just the central 2×2×2 cells.

Unlike for the octagonally symmetric case, the seed is now in 3D, however the computation remains linear on the number of pixels in the image because the limited parent neighbourhood allows the cells in most of the higher resolution grids to be ignored. The automata need only act in a small neighbourhood around the diagonal plane that intersects the 3D grid.

This method of increasing the rotational symmetry could be continued. We could start with a 4D grid of hypercubes and apply the rule to the nearest parents in 4D, then view just the 2D diagonal cross-section of the grid, the resulting automata should have octagonal symmetry. However, the search space would increase dramatically with each extra dimension used.

The results so far indicate that the automata with greater rotational symmetries do produce more organic and natural structures. So, with this in mind, there is a different way to achieve octagonal symmetry without these extra dimensions and the resulting large search space, they are based on the *diamond-square algorithm*, where *diamond* here means a square that is rotated by 45°.

The algorithm is so named because it takes a diamond lattice of pixels and fills the centre pixel of each diamond based on a rule on the four diamond pixels, then it does the same thing on the resulting square lattice of pixels:

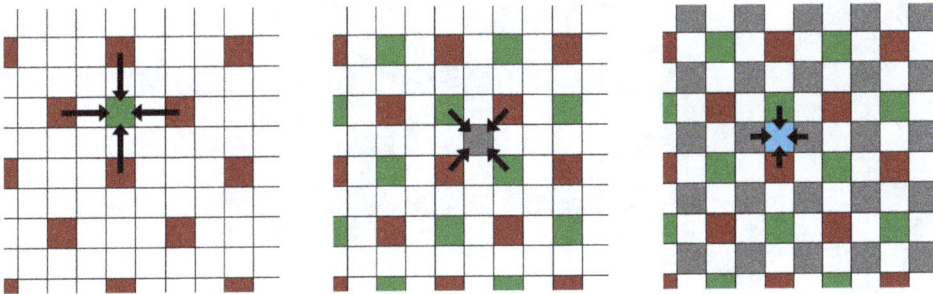

Figure 11.   Diamond-square algorithm repeatedly sets each cell from a rule on its four parent cells (arrow origins).

The process repeats until the entire grid of pixels is filled. These automata are different from the previous stack-of-grids approaches as they only use a single grid. Also, the parent cells are not just from the previous iteration but can be from any previous iteration, including the sparse initial starting lattice of pixels. This is seen in the above image as the differing colours of the parent pixels. The automata are nevertheless still scale-symmetric.

The geometry of the child to its four parents has the symmetry of a square, so this high degree of symmetry is used to minimise the search space. In fact, there are only six distinct permutations of 0 and 1 for the parent pixels, so even without employing bit symmetry, we have only 64 rulesets. This complete set can be viewed as an 8×8 collage.

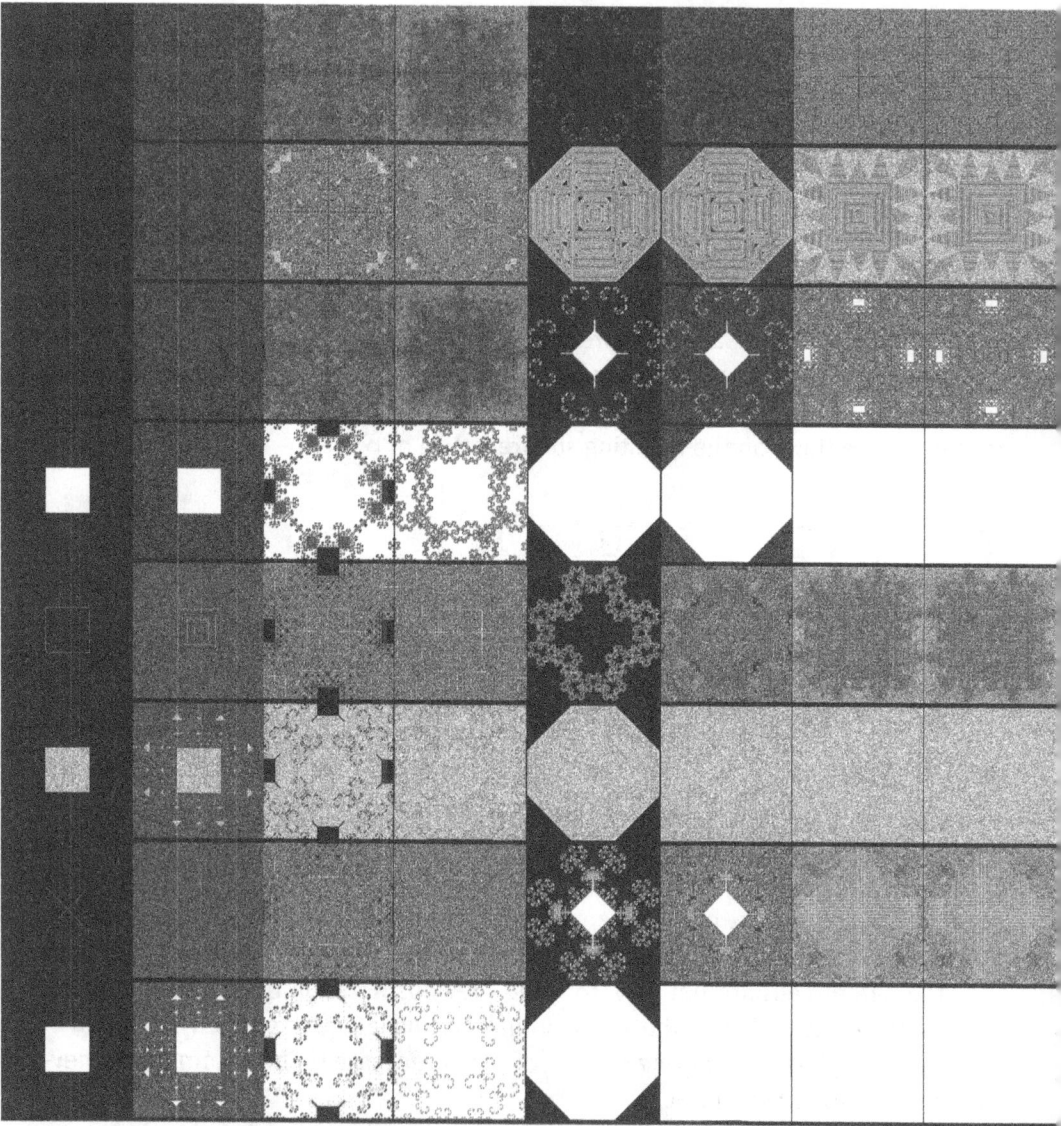

Figure 12.   Complete diamond-square automata ruleset using a 4×4 cell seed with just the central 2×2 cells set.

We cannot apply bit symmetry on these rulesets as there are an even number of parents. So, if two parents are 0 and two are 1, there is no choice of child state that inverts when the parent values invert.

Since the search space is so small, we can afford to explore a bit more extensively by increasing the number of states from two to three. Let us represent these states as the colours red, green, and blue. Due to the larger state space, we will reintroduce the bit symmetry constraint, which is now possible. We should call this *colour symmetry*, meaning the automaton is unaffected by swapping red, green and blue states in any permutation. It turns out that for three colours, there are now only three rulesets that have square symmetry *and* colour symmetry.

Figure 13.   Three diamond-square, colour-symmetric automata, using a 4×4 seed of red cells with a 2×2 green centre.

Figure 14.   Close-up of the third diamond-square automaton in the previous image.

Two of these rulesets give a dense mixture of the three colours. The third ruleset is perhaps more interesting, its rule for each new cell is simple: *choose the most common of the parent colours, if there is no single most common colour, then choose the least common*. This results in a scale-symmetric pattern. The pattern is entirely built of one shape that looks like a polygonal tear drop with small protrusions. These protrusions contain small child protrusions in a recursive fashion, and therefore, the shape is a tree, even though the child scale is so small as to not be easily noticeable. The tree shape is therefore a polygon with an infinite number of sides. These shapes form a scale-symmetric tiling of the plane.

Interestingly, there are no colour-symmetric rulesets for any greater number of colours, nor any in more dimensions than two, so these three types are unique. We are still able to change the lattice structure however, and in 2D, the other structure that has a scale-symmetric iteration is the triangular lattice, which can be visualised as a honeycomb.

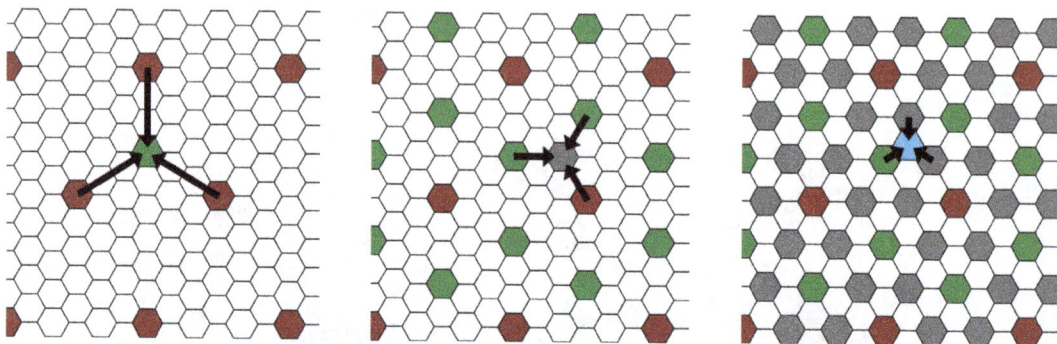

Figure 15.   Iteratively filling the triangular lattice automaton using three parent cells.

In this lattice, there are just two automata that have the triangular symmetry and colour symmetry, and they occur with exactly four colour states[1]:

---

[1] There are also three 2-colour automata with bit symmetry in this lattice. But one is a trivial case and the other two are the same automata as shown above, just a special case using only two colours in the seed points.

Figure 16.   The two triangle-lattice colour-symmetric automata using random seed colours.

Figure 17.   The second triangle-lattice automaton applied to a symmetric initial seed.

Again, one of the rulesets produces a dense mixture of colours and the other preserves the parent colours. The latter case is an attractive and unique automaton. Like its square lattice counterpart, it is built of solid polygonal shapes, and its rule is identical: for each new cell, *choose the most common of the parent colours, if there is no single most common colour then choose the least common*. But unlike the square grid case, this automaton also includes a shape that is a dense mixture of the four colours, and this shape has a fractal boundary very similar to the Koch snowflake. It is as though it has included a fifth colour. The overall appearance is like a beach of different coloured pebbles, where no two pebbles of the same colour ever touch.

Another extension we can apply to these upscaling automata is to generate them in three dimensions.[2] To do this, we will have to revert back to the stack of grids approach and use 3D grids for each layer. In fact, we have already generated 3D automata in the hexagonal symmetry examples, but we only rendered the long diagonal 2D cross-section. Each 2D image has a full 3D structure that can be rendered.

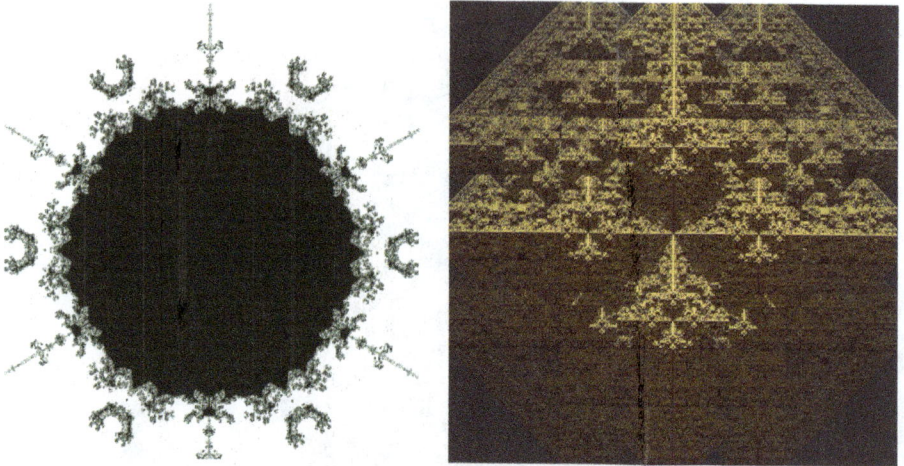

Figure 18.   Left: hexagonally symmetric automata. Right: its 3D structure.

Revealing the 3D structure is often quite difficult on a 2D screen, especially as fractal structures have no well-defined surface angles on which to perform shading. Here, we make use of shadowing and distance-based fog instead to provide a sense of depth.

---

[2] A form of 3D diamond-square method has also been previously explored [39, 40].

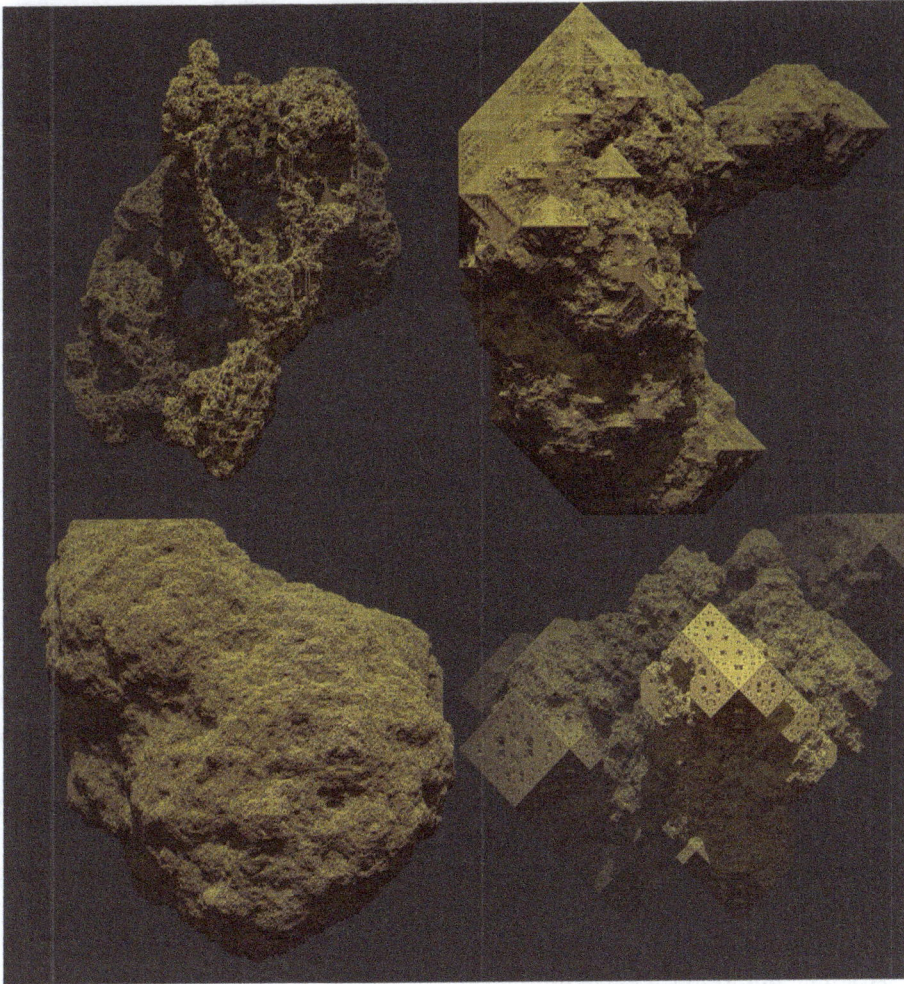

Figure 19.   Four 3D scale-symmetric automata with cube-symmetric rules.

As with the 2D case, these automata can have a variety of forms: sponge-like with holes through them, rock-like and shapes reminiscent of crystals or minerals that share the symmetry of the cube. In 3D in particular, it helps in understanding an automaton to also apply it to different initial seed grids of known shapes.

Figure 20.   Left: randomly seeded rock automaton. Middle: applied to an initial 4×4×4 C-shaped seed. Right: applied to an O-shaped seed.

While this study of automata has been an entirely geometric exercise, these 3D automata in particular could have practical application in rendering large 3D scenes. Imagine you wish to place a rock in the scene of a 3D program. The algorithm provides detail from a very course voxel seed to any level of detail required. So, the resolution can be calculated as needed based on the distance from the camera. Procedural rendering like this has low data bandwidth, so it can be rendered efficiently on graphics hardware.

Many different rocks can be built from coarse seed voxels and each can be finessed with user-defined details at any resolution. In addition, each rock can be split into two or partially destroyed by adjusting the low-resolution seed voxels, and the result will still have a rocky texture. One scene can use many different rulesets, with each low-resolution voxel storing the index of its ruleset. When two different rules are used on adjacent voxels, the automata tend to blend quite naturally at the boundary. So, a coarse world of voxels defined by their rulesets can be iterated into a high-resolution scene of complex, rough and connected 3D structures.

A simple addition that is helpful for real scenes is to treat the vertical axis separately to the horizontal axes and treat up as different from down. This is equivalent to giving the ruleset the symmetries of a pyramid rather than a cube.

Figure 21.   Top: rock and tree-like texture using pyramidal symmetry. Bottom: these two rules applied to the same coarse voxel mountainous scene. Detail is generated in proportion to proximity.

In addition to automata that add detail to a coarse grid, we could also produce an opposite sort of system that generates lower resolution from high detail grids. A useful example would be the process of smart image downscaling that retains important details of the high-resolution image. If you have an effective rule for downscaling an image to half its resolution, then the rule can be repeated at smaller and smaller resolutions, producing a stack of layers much like the upscaling automata do.

So, we have now introduced three general types of automata: those that take input from their neighbours (*cellular automata*), those that take input from their larger scale parents (*upscaling automata*), and those that take input from their higher resolution children (*downscaling automata*). By combining all three together, we can create animating *scale-symmetric automata*.

In this system, all three layers of inputs serve a purpose, the parent layer allows the automaton to act on the large-scale features of the image, the middle layer provides local interaction and the child layer allows the automaton to extract and act on the high detail features. For example, a crack may start as a fine detail, but the upwards information flow allows the larger bodies to eventually pull apart. The downwards flow allows the small cracks to move with the large bodies.

This is not to say that each of the three levels of input is essential, the middle layer does not seem to be required to ensure scale-symmetric automata that animate. However, we get most possibilities by utilising all three layers.

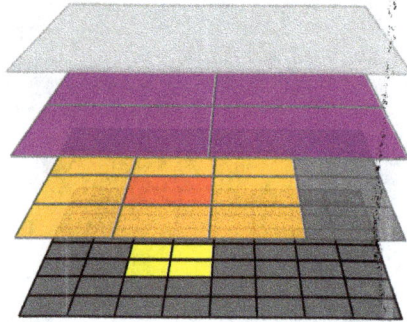

Figure 22.   Example system for a *scale-symmetric automaton*. The rule to choose the bit state of the central cell in red is based on the Moore neighbourhood of eight orange cells, the closest four larger-scale cells (purple) and the four half-scale cells (yellow) that it overlaps.

In order for the representations in each layer to stay synchronised, it is necessary for each layer to be updated with a frequency proportional to its resolution. If it did not, then a moving shape, such as a *glider* in Conway's Game of Life, would move too quickly at the larger scale, and there would be no gliding rule capable of keeping up with it at the smaller-scale levels. So, scaling the time dimension along with the spatial dimensions gives the same maximum speed for all the layers and keeps them physically in sync.

To accommodate these multiple update frequencies, we need to define the ordering of each layer's iteration. We can think of the ordering using a complete binary tree (where each branch has two children), each level below the root represents a smaller-scale grid size. Traversing the tree from left to right orders the iterations of each layer and maintains the desired relative frequencies.

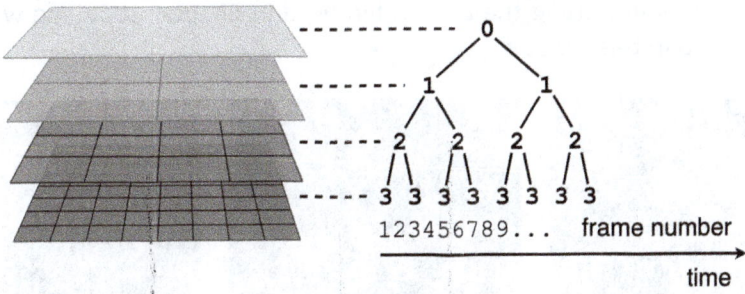

Figure 23.   Scale-symmetric automata update according to the left to right traversal of a binary tree.

This method has a surprisingly simple implementation: maintain an integer k starting at 1 and increment it with each time step. The position of the final 1 in the binary representation of k is the index of the layer to iterate.

Frame	in binary (final 1 is in red)	Layer to update (position of final 1)
1	0001	3
2	0010	2
3	0011	3
4	0100	1
5	0101	3
6	0110	2
7	0111	3
8	1000	0

This layer sequencing method updates each layer at a rate proportional to its resolution.

Each iteration for a layer just applies a ruleset to each cell in the layer. This ruleset is a function of the cell's nearest parent, neighbour and child cells, which generates a new 0 or 1 value for the cell.

As with the upscaling automata, we can search through the variety of results using the select-and-mutate user interface, based on random starting states for each ruleset, however this time each candidate is animating.

Also, like the upscaling automata, the results are highly varied. Many have strobe-like behaviours or quickly become saturated to either 1 or 0. Of the remainder, many act like moving amoebas and slowly coalescing blobs, others like spurting lava and flames. Some appear like a frothing sea, and generally, the behaviours act much more like a dynamic and organic system than traditional cellular automata do. Occasionally, there are results that retain the clear right-angled geometry of the system, creating growing lines for instance. These behaviours are viewable online [21] and can be generated using the code referenced in Chapter 10, along with all the other automata in this chapter.

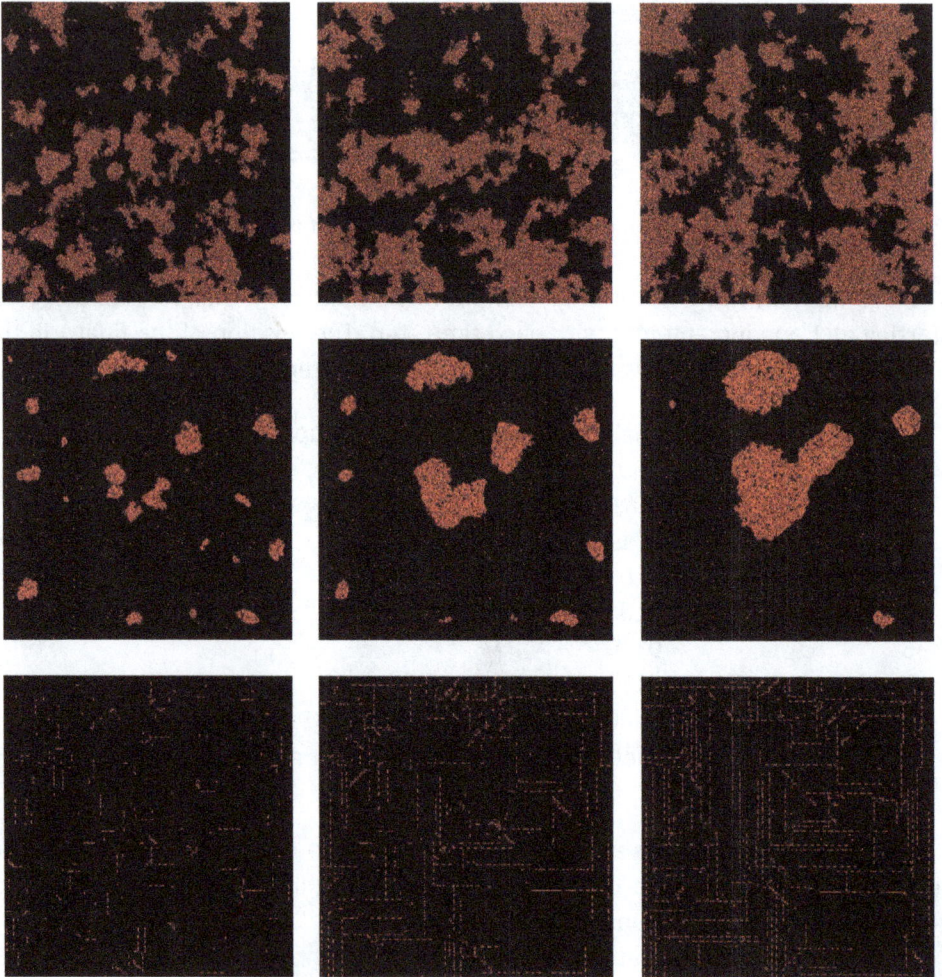

Figure 24.   Scale-symmetric automata examples. Top: lava-like behaviour. Middle: growing and combining blobs. Bottom: an unusual behaviour with growing dashed lines.

The basic system (Figure 22) has 16 closest neighbours plus the cell's own state as input. This represents a huge search space, far too large to usefully search using select-and-mutate. So, in order to reduce the search size, we can employ several constraints to filter out the uninteresting behaviours:

Spatial coherence: *If the nearest neighbours around a cell in a particular layer are predominantly 1 or predominantly 0, then set the cell to that value.*

This avoids behaviours that are noisy mixtures of 1s and 0s, and preferences solid shapes.

Temporal coherence: *If the cell's previous state was x, then the average output of the ruleset over all other neighbour values should be x.*

This avoids behaviours that strobe or jitter between 0 and 1 over time.

Layer coherence: *If the nearest parent and nearest children of a cell are predominantly x, then set the cell to value x.*

This ensures that each layer approximates the whole shape at its given resolution.

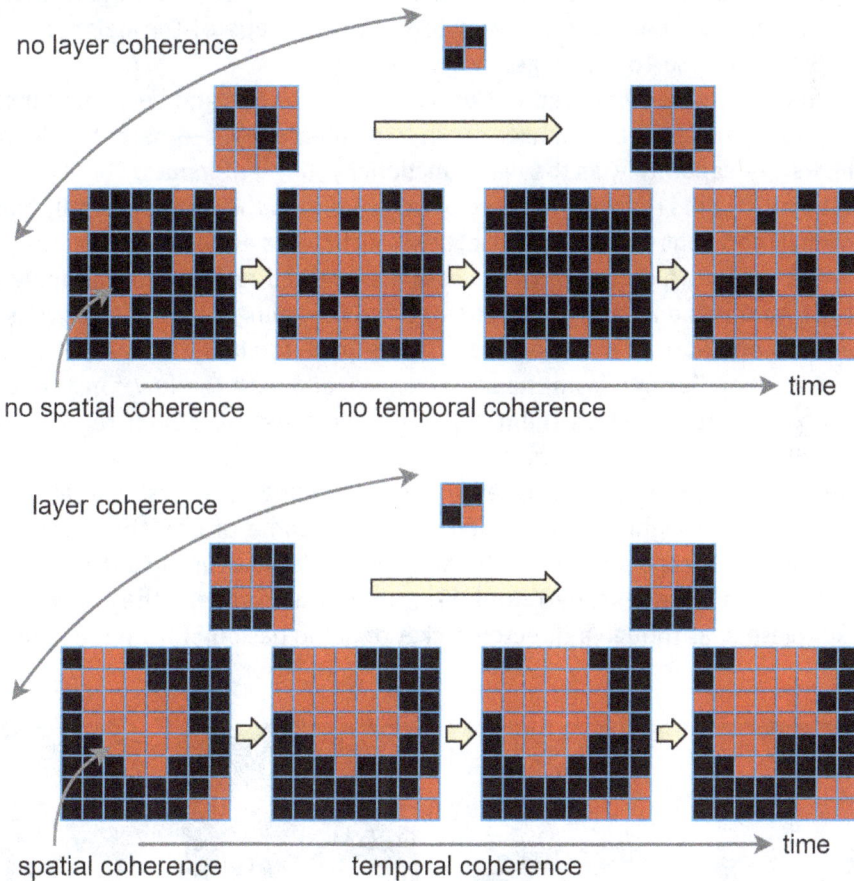

Figure 25.   Example scale-symmetric automata evolving over four frames. Top: without spatial, temporal or layer coherence. Bottom: with spatial, temporal and layer coherence.

We can also use additional symmetries to reduce the size of the search space:

Bit symmetry: *If the rule on the set of neighbours gives x, then the rule on the bit inversion of the neighbours gives 1 − x.*

This constrains the ruleset so that 0 and 1 (or black and white) are interchangeable without altering the behaviour.

We used this symmetry earlier. But even with this symmetry, the search space is still large, so another, more aggressive reduction is to use what I will call *shape symmetry*:

Shape symmetry: *Make the ruleset a function of the number of nearest parent, neighbours and child cell values rather than a function of each individual state.*

This makes the behaviour invariant to the exact shape of the neighbourhood, only the amount of surrounding geometry matters.

This shape symmetry is used in *Conway's Game of Life* and certainly simplifies the search space. However, it makes sense to sum each level — parent, neighbours and children — separately, as they are functionally different inputs.

These constraints reduce the space of behaviours to the most physically feasible ones, which represent scale-symmetric, dense structures evolving continuously over time. It isn't clear which are most necessary or indeed how strictly to apply each constraint, so this is an area that can be explored by trying different rule types with different combinations of the above constraints. With too many constraints interesting behaviours are lacking from the search space, and with too few constraints, the space is too big to ever find them. So, exploring these automata requires some experimenting with this trade-off.

Like with the Game of Life, scale-symmetric automata can also contain gliders, which move in a straight line. However, unlike the Game of Life, these will usually have a scale-symmetric structure. The simplest such structure in 2D is a triangle, moving forwards like a V-shaped front. We can find such gliders, they appear in the lava-like rulesets, as though a directed rocket of flame has burst from the main body of lava.

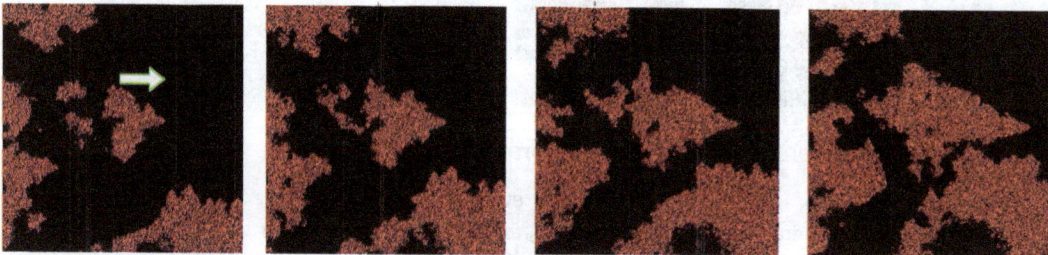

Figure 26.   Example of a fractal automata glider: moving from middle to right and growing.

Figure 27.   Triangle formation grows downwards in this different automaton.

However, usually these gliding structures have a varied and non-repetitive shape, often dying out before travelling too far. We might try and search for a precise and exactly repeating glider configuration for a particular ruleset, but I am not sure that this is a particularly useful feature to isolate in scale-symmetric automata, their behaviour is more irregular than the ordered mechanical systems that we see with standard cellular automata. The gliders that we see emerge, like those above, are ones that are robust to variations in the shape of the structure.

When simulating these automata, it is necessary to deal with the boundary conditions. These examples are all bounded within a square, so some cells lack neighbours on one or two sides. In addition, there is always a maximum scale top layer, which has no parent layer, and a high-resolution bottom layer, which has no layer of child cells. In all these cases, we could treat the missing neighbours as having value 0.

However, this adds a bias which interferes with the behaviour. To reduce the effect of the boundary, we could assign each missing neighbour's value to 0 or 1 with equal probability. Better still, if we employ spatial coherence, then we can treat missing neighbours around the cell as having the same value as the cell itself, and if we employ layer coherence, then we can treat the missing parent or child cells as having the same value as the cell. These are more accurate placeholders for the missing cells.

With the effect of the boundary minimised, the boundary is able to change without disturbing the simulation significantly. For example, if a new high-resolution layer is added, the influence from the parent cells will fill in the new layer, and within a few updates, its structure and behaviour will be consistent with the lower resolution layers. This sort of self-consistency within the structure is particularly the case when using the spatial, temporal and layer coherence constraints.

This robustness makes these automata well suited to simulating dynamics in large virtual worlds, where only a finite section of the world can be processed at any one time, and this section moves with the user.

In particular, it is well suited to first-person perspective worlds. In this perspective, there should be more detail and fidelity with nearby objects than distant objects. We can achieve this very simply: rather than using a stack of layers of the same width, we maintain a pyramid of layers of the same number of grid cells. In this pyramid, the largest scale cells also cover a larger area or volume of the world.

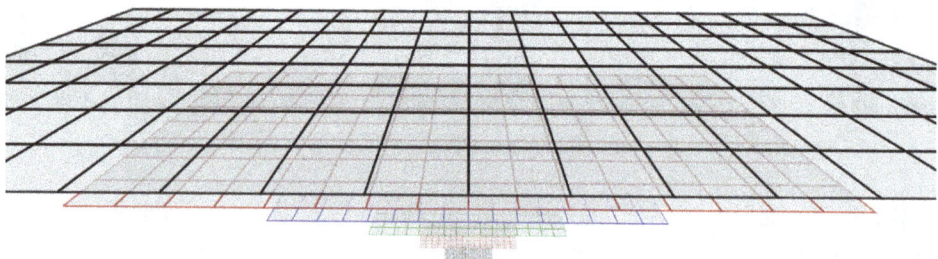

Figure 28.   Cross-section of the layers for a 2D world using 16×16 cell grids centred around the user.

The pyramid doesn't slide with the user's movements, as each cell represents a fixed location in the world. Instead, new rows of grid cells are added and removed in order to maintain the pyramid shape around the user. The new cells' values are then a function of their larger parent cells and their immediate neighbour cells. They will then converge to a value that is compatible with these cells. So, this system can maintain local consistency, giving the impression of a globally consistent simulation, without storing and processing the entire world in high detail.

Consequently, the computational complexity of this design is very low. The memory usage for a world of radius r is $O(\log_2 r)$, and the processing cost and data bandwidth are just $O(1)$ with respect to r due to the fact that larger scale layers are updated less frequently.

We could exploit this self-consistency property a bit further. In addition to allowing the user to move spatially, we could allow the user to zoom into or out of the world, in other words, shift down or up the pyramid of layers. This allows an unbounded range of scales using a finite number of layers.

These moving bounds approaches work best for the organic style of automata seen so far, like sloshing lava, growing blobs or changing clouds, where the user will not notice global inconsistencies. But the underlying system can be applied to much more general simulations as well. We could generalise the ruleset to operate on arbitrary data structures, rather than bits, and to use any arbitrary function of these inputs.

For example, a cloth simulation could represent a square patch of the cloth per grid cell and store the patch's position and velocity as the cell's state. The ruleset would then be a function that applied the local cloth dynamics. It would be interesting to see how well these multiscale automata work in more physical situations such as this.

All these different ideas illustrate that *scale-symmetric automata* is a topic with many potential applications and much left to discover. In addition, its underlying hierarchical update system looks to be particularly effective in immersive world environments.

★

We have now explored the idea of applying scale symmetry to automata. Compared to traditional cellular automata, the images and behaviours appear more organic and somewhat more akin to physical processes. The overall update scheme, whereby the update period is proportional to the scale, seems to be a useful scheme in a lot of settings, and we have considered some possible applications. In the next chapter, we will look more directly at physical processes and will see how scale symmetry applies under more physically realistic dynamics.

Chapter 7

Dynamics and Physics

The automata in Chapter 6 followed a rule whereby the smaller pieces of the automata operated in a proportionally smaller timescale. This means that patterns of shapes and movements were scale-symmetric in space as well as in time, we could say that they have "spacetime scale symmetry." This symmetry is not unique to automata, we can apply it to other dynamical systems too. Usually, we refer to n space dimensions and the single time dimension as $n + 1$D spacetime, so let us investigate these $n + 1$D scale-symmetric systems in this chapter.

There are many examples of these in nature, such as swaying trees, waves on water and colliding asteroids. We will examine these later, but let's start with a simple example of a set in 1+1D which represents a row of repeatedly colliding segments:

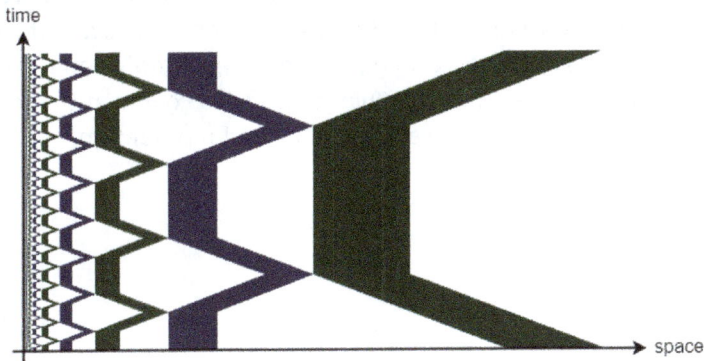

Figure 1.   Time sequence of a set of colliding line segments shows symmetry of scale in space and time.

This set represents a *kinematic* motion as it does not include any consideration of masses and forces. As such, it is unlikely to represent any mechanism in nature. To represent natural mechanisms, we need to include mass and so look at scale-symmetric *dynamics*. To do so, we need to ensure that the set represents objects that obey the basic laws of motion.

If we give all the segments in the above set equal density, then they each have a mass and we should expect their collisions to conserve momentum and energy. In fact, this example does conserve momentum at each collision, but the collisions do not conserve energy, so it does not represent a valid dynamical system of rigid bodies.

It is not immediately obvious that it is possible for scale-symmetric systems to be physically valid, but we have some basic evidence to encourage us. The simplest implementations of $n + 1D$ scale symmetry have all parts moving at the same maximum speed, which means that structures of constant density will have a finite kinetic energy in any finite volume. What's more, no parts will exceed the speed of light if the base part doesn't. So, physically valid systems seem to at least be plausible.

The following examples show that such systems are not just plausible, they are in fact commonplace.

Billiards

We can extend the previous example beyond 1D to generate a system of colliding objects that follow a size power law. This behaviour is like the collision of pebbles in a churning current or of colliding asteroids in the asteroid belt.

We could call the mathematical description of this *scale-symmetric billiards*, whereby the ball sizes follow a power law.

We can solve this in 2D with disks of constant density distributed in a tree structure, such that every disk of radius r has two child disks of radius $r/2$. The base trajectory of the tree consists of two disks of equal radius repeatedly bouncing together in symmetrical v-shaped paths. The two child trajectories collide at the locus of these two v shapes in a collision that conserves energy and momentum.

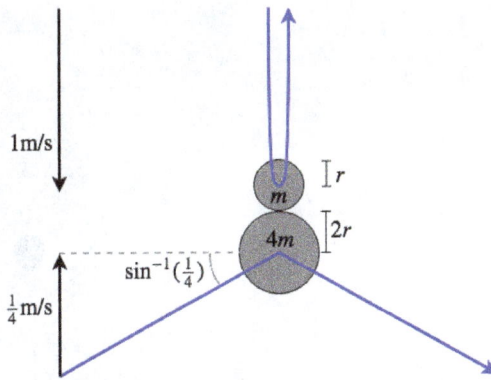

Figure 2.   Frictionless elastic collision between disks travelling at equal speed, configured to conserve momentum and energy.

This physical requirement constrains the angle of the v to 2acos(¼) for disks travelling at equal speeds. The equivalent of the billiard table edges is a set of fixed disks that the moving disks can deflect off, these could be joined up as a continuous boundary surface if required.

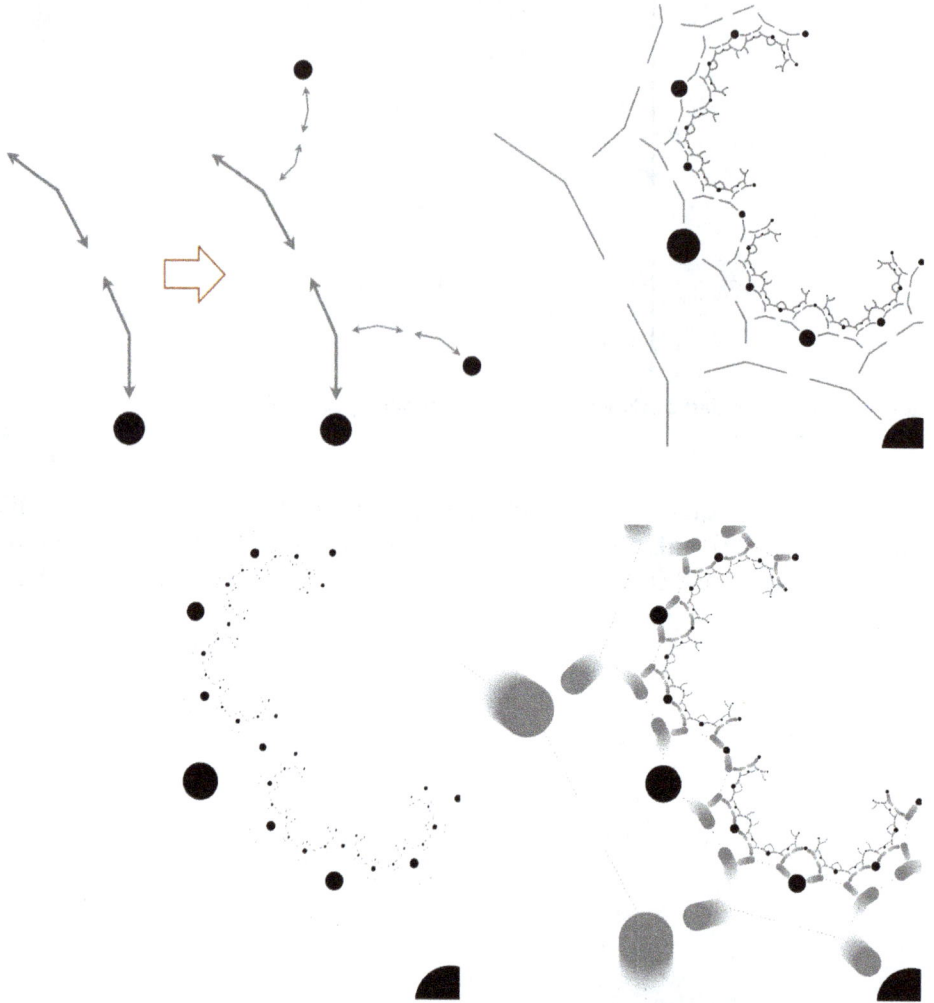

Figure 3.   Top: base path of two oscillating motions in grey, and first children. Right: all children. Bottom: the pattern of fixed disks. Right: the disks moving along the paths over a short time interval.

Since there are two children for each parent, of one quarter the mass each, the mass in each generation decreases exponentially, while the disk speed remains constant. This means that the total mass and the total kinetic energy in the system are always finite for any finite sized base disks.

The mechanism itself is an unusual one, the largest disk's path has a free end, so must receive intermittent impulses to maintain the repeated bouncing behaviour. These localised impulses become more frequent and spread out as they transfer to the smaller and smaller disks. At the limit, they become a continuous pressure — in Newtons per metre — along the string of smallest fixed disks. It therefore transforms an input that is localised in space and time into an output that is diffuse in space and time.

There are lots of topics to explore here. It would be interesting to simulate and listen to the sound made by such a billiard table. Could a three-dimensional version be constructed, and would it sound the same? And what about the fixed disks, is there a construction that doesn't require them? If so, this would make a neat toy model for asteroid belt dynamics.

Sea

The motion of water is of course a complicated system to simulate, however its surface can often be well approximated by linear wave equations. This is particularly the case when the water depth is small compared to the wave length and when the wave height is also small. Under these conditions, the surface dynamics can be linearised as the sum of multiple sinusoidal planar waves, each travelling at the same speed [22]. We then require a model to describe the wave heights.

Wind-driven waves follow a spectrum that can be used to correlate the amplitude of waves with their wavelength, there are multiple available spectra depending on the sea and wind conditions, which have been found using real-world measurements [23]. In the implementation panel, we show that in shallow water conditions, the smaller waves appear to have amplitude approximately in proportion to their wavelength. Due to the constant wave speed, this wave model is therefore invariant to scaling space and time by any constant, so the dynamics have spacetime scale symmetry.

A simple implementation of this model uses a sequence of exponentially decreasing wavelengths and amplitudes of random orientation. As such, it is an example of a moving structure that is scale-symmetric in the statistical rather than absolute sense.

Figure 4.   Scale-symmetric wave dynamics. Here, we only include wave directions within one quadrant of the horizontal plane to give the suggestion of directed, wind-powered waves.

Sea Surface Implementation

We define a single wave function as:

$$y_i(x) = a_i \sin((x/v - t)f_i)$$

where a_i is the wave amplitude, f_i is the wave angular frequency and

$$v = \sqrt{gh}$$

is the wave speed, where g is the acceleration of gravity and h is the water depth. We use a constant depth, giving constant wave speeds.

The 2D wave surface is then defined for any horizontal position vector $\boldsymbol{p}$:

$$y(\boldsymbol{p}) = \sum_i y_i(\boldsymbol{d}_i \cdot \boldsymbol{p})$$

for a set of randomly distributed wave direction vectors $\boldsymbol{d}_i$.

In order to be scale-symmetric, the wave frequency of each wave i should be a constant multiple of the previous (we use $f_i = 2^i$) and also the wave amplitude should be proportional to the wavelength λ_i.

The surface is a sum of sine waves in random directions d_i, with amplitude and wavelength halving with each wave.

Since $\lambda_i = \frac{2\pi v}{f_i}$, this means that we require $a_i \propto f_i^{-1}$. We can justify this relation using wind-driven wave spectra:

In wave spectrum models [23], sea waves have a peak amplitude at a certain frequency that decreases with greater wind speeds. For waves with frequencies higher than this peak frequency, the spectrum tends to a simple power law. In shallow water conditions, it has been reported that the spectral density is inversely proportional to the wave's angular frequency cubed [24]:

$$S(f) \propto f^{-3}$$

The spectral density is a power per frequency band. In our model, we pick a single wave to represent each frequency band. So, if each wave i has frequency $f_i = 2^i$, then it encompasses a frequency band of width proportional to f and its power is therefore proportional to f_i^{-2} and the wave amplitude is the square root of this:

$$a_i \propto f_i^{-1}$$

and so the system is scale-symmetric.

Trees

The two previous examples use absolute velocities, and as such, the speed of every component is identical. Often, the child components act relatively to the coordinate frame of their parent, swaying trees are one such example. We can model these by using a sinusoidally varying branch angle on any scale-symmetric tree structure. The period of each sinusoid is then set in proportion to the length of each branch.

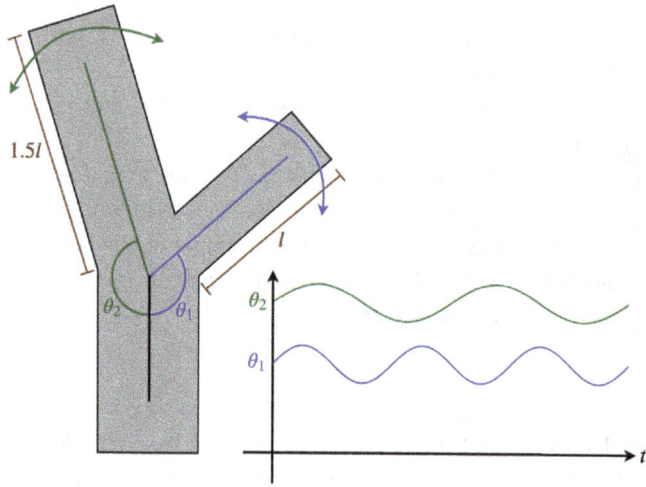

Figure 5.   Scale-symmetric swaying motion. The larger, green branch of length $1.5l$ sways with a proportionally longer wave period, taking 1.5 times longer to oscillate than the shorter blue branch. This proportionality to branch length is applied to every branch in the tree.

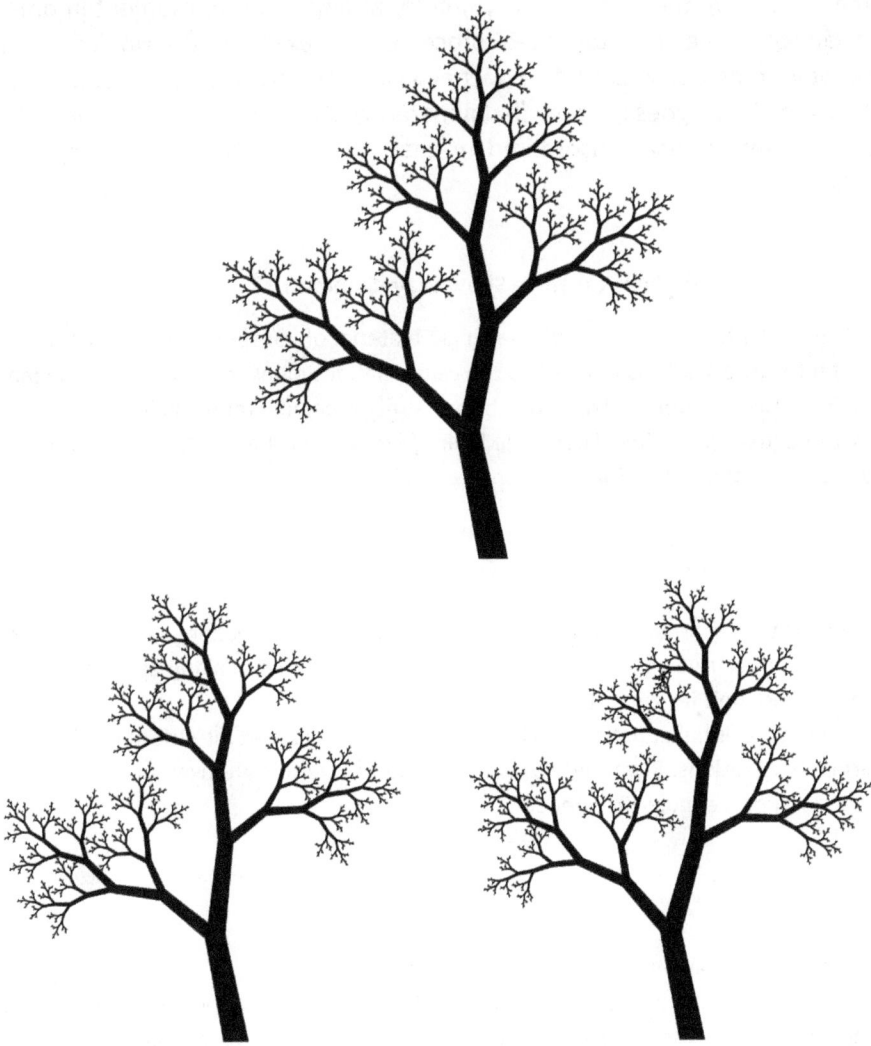

Figure 6.   Top: tree at time = 0 looks symmetric and artificial. Bottom: at time > 0, the visual symmetry is lost.

The scale-symmetric motion results in the smallest branches oscillating very quickly, which can be seen in real trees. We sometimes refer to this as the tree bristling or rustling in the wind. This non-smooth motion is characteristic of most space-time scale symmetry.

When the recursive construction uses differing scales in its branches, the tree's pose appears asymmetric and unsynchronised at almost any point in time, even though the pose at time zero reveals its geometric, artificial construction. The same is true of other dynamic examples such as the water surface. So, spacetime scale symmetry can give the illusion of asymmetry at any arbitrary moment in time. The same can sometimes be true of real processes. For example, the ripples on a pond may appear to be chaotic but do in fact align at time zero when, for instance, a rock hit the water. It just goes to show that apparently disordered structures do not imply that the system is itself disordered or that the scale symmetry is imprecise or stochastic.

Do Physical Trees Show Scale-Symmetric Motion?

If we model a tree as composed of a single material of constant density, then it has a constant *Elastic Modulus E*, which is the value that describes the stiffness of the material. If we then model the tree branches as straight beams, then small deflections follow the equations of Euler–Bernoulli beam theory. In particular, the free vibration of each branch follows the Euler–Lagrange equation:

$$EI\frac{\partial^4 w}{\partial x^4} = -\mu\frac{\partial^2 w}{\partial t^2}$$

where I is the second moment of area of the beam's cross-section, x is position along the beam, the curve $w(x)$ describes the deflection of the beam along its length and μ is the mass per unit length of the beam.

This differential equation can be solved and decomposed into the sum of multiple harmonic vibrations. Regardless of whether the beam ends are fixed or free, the principle vibration's angular frequency can be found as:

$$\omega = \frac{a}{L^2}\sqrt{\frac{EI}{\mu}}$$

where L is the beam length and a is a constant.

For any proportionally sized beam of length L and constant density, the mass per length is $\mu \propto L^3/L$. Also, the second moment of area is $I \propto L^4$ by its definition, so plugging these into the above equation, we get:

$$\omega \propto L^{-1}$$

The branch sway frequency is inversely proportional to its length. This is precisely the expected behaviour of 3+1D scale symmetry.

The added effects from child and parent branches make the precise swaying motion more complicated, but the scale symmetry remains, and a single sinusoid approximates the branch motion to first order.

Planets

One case that is often considered to violate 3+1D symmetry is planetary dynamics. It is true that if you double the dimensions of our solar system and maintain the density, the planets will not be orbiting with twice the orbital period of ours. But while the constant density criterion makes sense for earthly systems like water and tree branches, it isn't obvious that it is appropriate for the physics of gravitationally attracting bodies. In fact, the requirement for 3+1D scale symmetry says nothing about how masses should be scaled, since it is a purely geometric requirement.

We can however turn the problem of orbiting bodies into a geometric one by replacing each planet with a small black hole, the simplest model being the *Schwarzschild black hole*. The mass of each black hole is directly proportional to its radius, the Schwarzschild radius. It turns out that scaling up a system of black holes together with their radii does equate to a scaling of the orbital periods by the same factor, it is indeed scale-symmetric in spacetime.

While this replacement may seem like a rather extreme measure, the force of gravity from black holes far from their event horizon tends to that of a planet of the same mass, so we can still use simple Newtonian physics to model it. Moreover, we could replace each atom of the original planet with a tiny black hole if we wished, and the scale symmetry would remain, while maintaining the mass distribution of the planets.

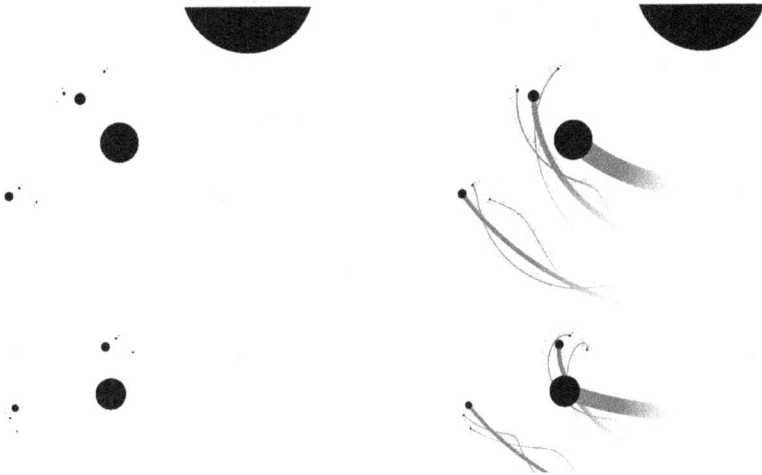

Figure 7.   An example black hole system at time 315 s. Right: tracing the paths over time.

Generating a Black Hole System

Here, we model a black hole system of recursive satellites. For each parent of radius r, the satellites follow a circular path with angle $\theta = t/T$ for some time t and orbital period T.

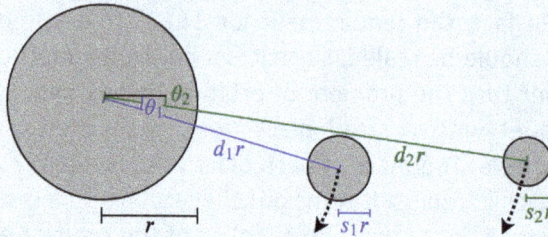

Using the *Schwarzschild black hole* model, we know its Schwarzschild radius is:

$$r = 2GM/c^2$$

where G is the gravitational constant, c is the speed of light and M is its mass. Since mass and radius are proportional, we can say that each child satellite of radius $s_i r$ has mass $s_i M$.

When d_i is large and s_i is small, the dynamics can be well approximated by circular motion of the satellite relative to the parent. From Kepler's laws, we know that the orbital period of a satellite with mass $s_i M$ follows the equation:

$$T^2 = (d_i r)^3 \frac{4\pi^2}{G(M + s_i M)}$$

Substituting in the Schwarzschild radius equation, we get:

$$T^2 = d_i^3 r^2 \frac{8\pi^2}{c^2 (1 + s_i)}$$

So, the orbital period T is proportional to the scale r, and therefore, the system has 3+1D scale symmetry.

To generate the animating system, we can place each satellite based on its orbital angle, using simplified units, this is:

$$\theta_i = \frac{t}{r}\sqrt{(1 + s_i)/d_i^3}$$

Each satellite then becomes the parent black hole to a smaller copy of the same system, producing a dynamic scale-symmetric solar system.

★

The examples so far show how structures with spacetime scale symmetry can also be physically realistic and represent approximations of well-known physical processes. Indeed, studies have found this symmetry to be present in many other natural phenomena from boiling and dripping water to heart beats and rural soundscapes [25, 26].

Given this prevalence, it seems plausible that many of the fractal shapes we recognise in nature are actually built from $n + 1D$ scale-symmetric systems that are just evolving slowly, for instance, the generation of clouds and mountains, the growth patterns of forests and the evolution of rivers. It would be interesting to investigate the power laws involved with these processes and see if this is indeed the case.

A problem with the given tree and planet examples is that the child parts act in the local space of their parents and consequently the velocities in a chosen world frame are unbounded and can exceed the speed of light. Of course, this may not be of great concern to a geometrist, but since we are progressing towards more physical models, it is worth considering further:

Is it possible to have a recursive, scale-symmetric system that allows relative motion without breaking the light barrier?

The answer is yes, but to do so, we need to use a very different coordinate system. Rather than the Newtonian spacetime of Isaac Newton's laws, we need to use the more mind-bending *Minkowski spacetime* of Einstein's relativistic laws. Fortunately, this is not quite as complicated as it sounds.

If we draw a coordinate frame with a vertical time arrow and a horizontal space arrow such that light speed is at a 45° angle, then we can call this our world coordinate frame. In this frame, a stationary point is a vertical line, and if this point moves leftwards, then the vertical line leans to the left, if it moves rightwards, then the line leans to the right.

In Newton's laws, shifting to the coordinates of someone travelling leftwards is represented as a rightwards horizontal shear of everything in the world coordinate frame. This shear preserves areas on the grid and is named the Galilean boost.

The same velocity change in Einstein's laws (*Minkowski spacetime*) is a stretch in one diagonal axis and a compression in the opposite diagonal axis. This transformation also preserves area on the grid and is named the Lorentz boost.

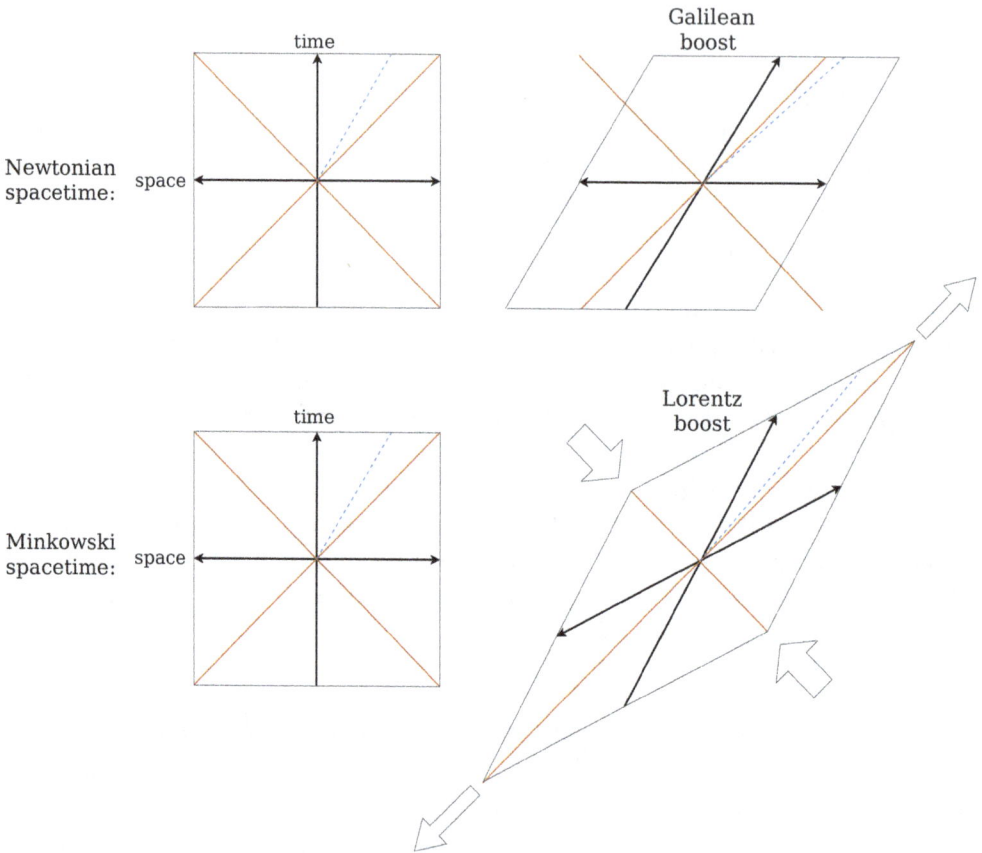

Figure 8.   Top: the Galilean boost of standard Newtonian dynamics causes a moving object (blue dashed line) to exceed the speed of light (orange lines). Bottom: the Lorentz boost stretches and compresses spacetime along the light speed axes, so nothing crosses these limits.

While repeated shearing will send an initially vertical line beyond the 45° angle, the repeated diagonal stretch cannot cause lines to cross 45° and exceed the speed of light. This Lorentz boost is the mechanism that forms the basis for Einstein's Special Relativity and gives a geometric reason why light speed is never exceeded.

So, let's use this Lorentz boost to generate a scale-symmetric trajectory that also remains within the speed of light. We base the example on the Koch curve, but in the 1+1D *Minkowski spacetime* rather than in 2D space. This represents the trajectory of a point moving in 1D space, rather than a spatial curve. Its generation is straightforward:

We start with a vertical vector (representing a stationary point). Then, for each iteration, we split each vector into two by taking its mid-point and offsetting it by a portion *k* along the orthogonal vector. This is just like a Koch curve iteration, but in 1+1D spacetime, the orthogonal version of a vector (*x,t*) is (*t,x*) rather than (*t,−x*).

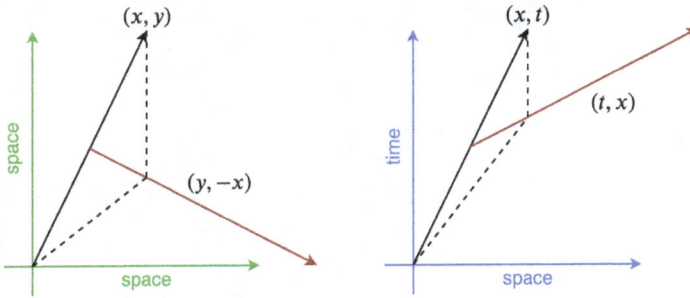

Figure 9.   Left: for the standard Koch curve, we offset the mid-point of each segment along the orthogonal vector (in this case, one quarter of the way, so *k* = ¼). Right: in 1+1D spacetime, the orthogonal vector has the opposite sign in the vertical axis.

As with the Koch curve, we repeat the process, swapping the shift direction each iteration. This generates a sequence of increasingly complex trajectories:

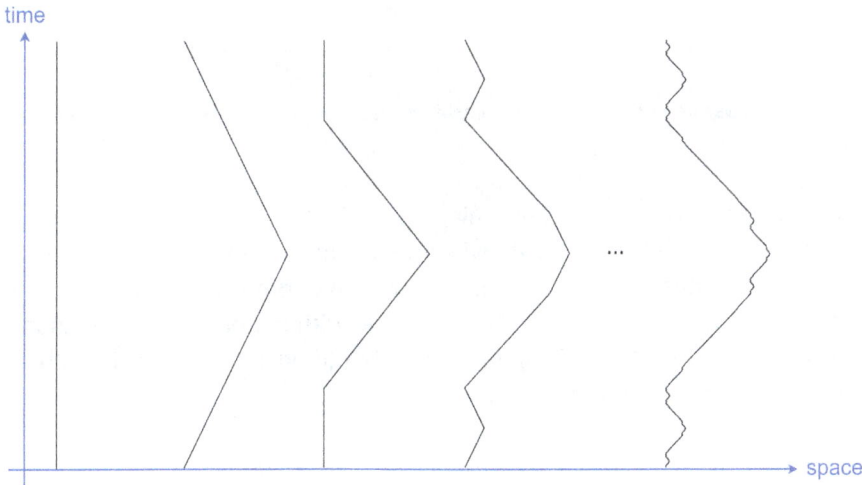

Figure 10.   The first four iterations on the Minkowskian Koch curve and the final iteration.

We can measure the duration of a clock on the trajectory as the sum of $\sqrt{\delta t^2 - \delta x^2}$ for each small interval $(\delta x, \delta t)$. This is called the proper time interval of the trajectory. This allows us to use the *yardstick method* from the introduction chapter to calculate the fractal dimension of the trajectory. Interestingly, the total proper time *decreases* with every iteration, unlike with the spatial Koch curve, where the length only gets longer. A consequence of this is that the fractal dimension of the trajectory actually reduces for larger bend factor k rather than increasing.

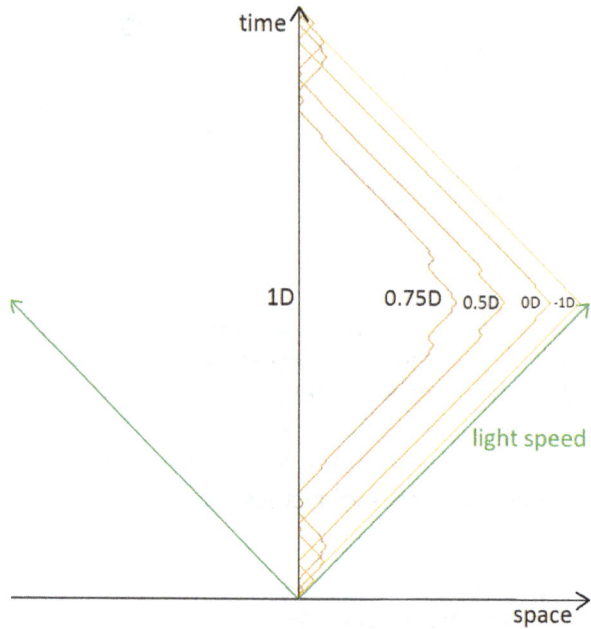

Figure 11.   Minkowskian Koch curve for increasing offset k, which corresponds to decreasing fractal dimension.

This is also different from Newtonian spacetime where the time period of the trajectory is unaffected by the subdivision, causing the trajectory to remain one-dimensional regardless of bend factor. In fact, these three cases form a neat triplet of scale-symmetric structures: the 2D Koch curve with dimension greater than 1, the 1+1D Newtonian (*Weierstrass*-like) function with dimension equal to 1 and the 1+1D Minkowskian Koch with dimension less than 1.

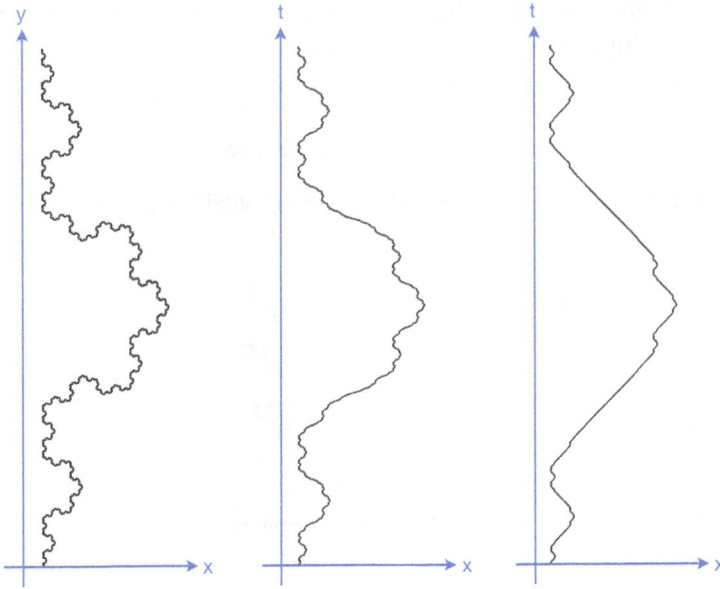

Figure 12.   Sibling fractal curves. Left: Koch curve in 2D space. Middle: *Weierstrass*-like function in 1+1D Newtonian spacetime. Right: the equivalent curve in 1+1D *Minkowski spacetime*. Interval lengths are $\sqrt{\delta y^2 + \delta x^2}$, $\sqrt{\delta t^2}$ and $\sqrt{\delta t^2 - \delta x^2}$ respectively, where the latter two represent the time interval for the moving point.

An unusual property of this Minkowskian trajectory is that its temporal dimension can drop to zero and even below zero with no lower bound. There is nevertheless a logical justification for the behaviour, it means that the measured proper time interval of the trajectory reduces as you measure it at higher resolution. The higher resolution reveals that the trajectory is covering more ground, the higher velocity means clocks on the trajectory tick more slowly. This roughness in velocity is not really evident in the graphs, on the contrary the velocity profile appears to get smoother with increasing offset. But this is just a drawback of the way it is depicted, the velocity variation is not noticeable when close to the speed of light, but if we boost our coordinate frame into that of the fast-moving point, then the roughness reveals itself.

Just as the Koch curve has discrete rotational symmetry, this Minkowskian Koch curve now has discrete boost symmetry. That means that the same trajectory shape can be found if the observer's coordinate frame is boosted any integer number of times. So, the trajectory has no special velocity that it is relative to, which we could call stationary. In shorthand, we could refer to this trajectory as *relativistic*.

One step further is to look at 2+1D Minkowski spacetime. Here, we generate an analogue of the 2D Levy C curve. We start with a trajectory vector (x,y,t) and repeatedly split it into two, offsetting the mid-point a distance d orthogonally to the vector's spatial direction. The distance d is chosen so that the split trajectory's proper time is multiplied by a constant factor k.

2+1D Levy C Curve Iteration

Every trajectory vector (x,y,t) is split into two (the dashed lines) through a new vertex v. Shown here in the x,y plane:

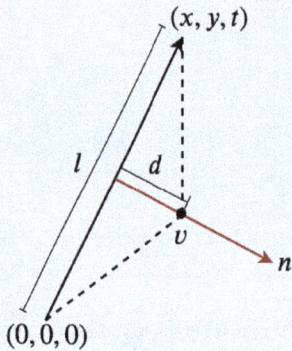

length: $l = \sqrt{x^2 + y^2}$

speed: $s = l/t$

offset distance: $d = \frac{t}{2}\sqrt{\left(1 - k^2\right)\left(1 - s^2\right)}$

normal direction: $n = (y,-x,0)/l$

new vertex: $v = (x,y,t)/2 + dn$

(x, y, t)

l

d

v

n

$(0, 0, 0)$

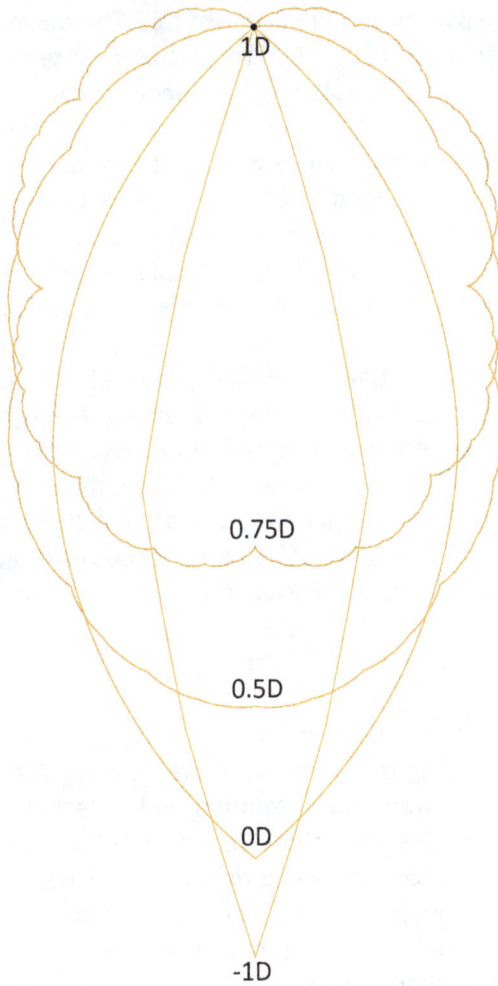

Figure 13.   2+1D Minkowski Levy C trajectories of reducing fractal dimension. The 1D trajectory is a stationary point.

Like with the trajectory in 1+1D, a 1D trajectory corresponds to a stationary point in a chosen world frame. For slightly smaller fractal dimension, the trajectory follows a shape similar to a Levy C curve, but with the bend angle reducing at smaller scales. As such, its spatial 2D curve is non-fractal and has a finite length. It is interesting to note that the projected curve gets smoother with reducing temporal dimension, and when the dimension is zero, the curve appears to be made of two circular arcs.

In fact, they are not quite circular, but as we increase resolution and the trajectory approaches light speed, the path tends to a circular curve. In this case, when the span of the arc is cos(30), the curvature tends to $\sqrt{3}$. It is only when the dimension is exactly zero that the path has a curvature like this. The converse may also be true, it suggests that any light-like circular arc through space has temporal dimension zero. I am not sure whether this result has any significance, but it certainly seems interesting.

As with the 1+1D case, one can extend the trajectory into negative temporal dimensions where the curve bend angle now appears to reduce too quickly for a smooth result and so has a convex, angular appearance. This curious case of a negative dimensional structures is specific to examples in Minkowski spacetime. We won't investigate it further here, but we will revisit negative dimensions in Chapter 9 in a more general setting.

These trajectories show that it is indeed possible to have scale-symmetric systems that allow relative motion without breaking the light barrier. They also showcase an extra fundamental symmetry that we have not used in previous chapters: boost symmetry. It would be interesting to follow this path further and develop versions of the waving tree and planetary bodies that don't exceed the speed of light. Not only is this more realistic, but bounded speeds make interacting with these structures a more computationally feasible prospect.

★

As you can see, we are heading more and more towards physically realistic settings, but what is guiding the route is symmetry. In addition to pure scale symmetry, we have included translation, rotation and reflection symmetry, and now boost symmetry. Symmetries are interesting because they reduce the information content of the structures. For instance, a right-angled triangle needs a full angle to define its orientation, a square only needs an angle between 0° and 90°, and a circle needs no angle parameter at all. This reduction means the structure has fewer degrees of freedom and is more unique.

This is perhaps the reason why symmetries are so common in physical laws, they contain less information in need of explaining. For instance, laws that are translation symmetric do not need any absolute position parameter, so positions are only relative to other positions. When this is the case, we say that absolute position is an unobservable quantity in the resulting universe. This certainly appears to be the case in our universe, and it is a core assumption underlying Newton's and Einstein's laws of physics.

We could continue this progression towards physical realism by including other known symmetries of nature. Taking the example of the scale-symmetric automata of Chapter 6, if each cell had a real-value representing electrical charge, then it would be interesting to experiment with *CPT symmetry*. This is a known symmetry on the product of charge (*C*), parity/handedness (*P*) and time direction (*T*) of particles. If each cell was a complex number, then it would be interesting to make the rulesets invariant to changes in the number's angle relative to the real line, mimicking the real-world *phase invariance* in the wave function of charged particles. There are many such paths this exploration could continue along.

In general, this is progressing towards a process of defining a set of physical laws and exploring the resulting universes that emerge. The symmetries keep the size of the state space low, but it still remains a vast and open topic of study.

Having looked into a wide range of examples now, from 2D shapes through to interactive and physical systems, we can now investigate categorising these specimens, which lets us observe how the structures relate to each other. This will be the topic of the next chapter.

Chapter 8

Classification

The last six chapters have explored a wide range of scale-symmetric algorithms. The resulting images are diverse and each type that we have looked at seems to have its own personal quality. Some are precise and regular, others wispy and moody. Some are grand and sweeping, others more modest and delicate. Many resemble natural phenomena, such as storms, lava, bubbles, rocks, mountains, cracks and trees. Others are artificial in appearance. It is easy to get lost, so what we need is to organise and parameterise these examples, to chart the landscape and so get a bird's-eye view of where the interesting and unexplored areas are. That will be the topic of this chapter.

Euclidean structures have many such categorisations, there are the regular solids, the polyhedra, the convex and concave shapes. Structures can be described by their topology, by their reflective and translational symmetries, and by predefined shape labels such as square, oval, cuboid, etc. But what systems do we have for classifying scale-symmetric geometry?

Certainly, there are some. For instance, the examples in this book are already categorised by chapter topic and then by algorithm. More generally, there are many well-known fractal methods, such as the *escape-time algorithm*, *substitution rules*, *Iterated Function Systems*, *Fractal Flames* and Kleinian *limit sets,* to name a few. However, all of these describe the generation method rather than the result. This is not the best way to categorise the resulting structures because one structure can result from many different generation methods. For example, the Sierpinski triangle can be generated with an Iterated Function System, as a *substitution rule* and even from a stochastic game, called *the chaos game* [27].

159

We could of course just label different interesting shapes, such as the *Sierpinski triangle* and *Menger sponge*. However, that doesn't help in classifying the diversity that we see. A helpful categorisation should be unconcerned by the exact shape of the structure, or which method generated it, but instead care about structural differences. What are the structurally different shapes?

The final topic in Chapter 1 provided an excellent foundation of an answer by showing that its simple 3×3 grid substitution scheme generates only five qualitatively different fractal structures in 2D. By expanding the domain of these types from pure fractal shapes to general scale-symmetric structures, we can describe five general classes in 2D, they are as follows: *void, cluster, tree, sponge* and *solid*. Labelled classes 1–5.

Void and solid are the rather trivial cases of empty space and filled space, respectively. In between, we have clusters, trees, and sponges. Clusters are scale-symmetric distributions of a base shape, trees are non-looping scale-symmetric distributions of branches, and sponges are scale-symmetric self-connecting networks.

(1)   Void: contains region that is empty of points and is continuously scale symmetric   —

(2)   Cluster: a recursive, disconnected set of regions   convex

(3)   Tree: single connected structure with recursively branching limbs   convex and concave

(4)   Sponge: single recursive network of connected limbs   concave

(5)   Solid: volume or area of space is filled, contains no finite sized gaps   —

Classification list for 2D scale-symmetric structures: images, definitions and expected convexity.

This list has a nice symmetry to it, if you reverse the list and swap black with white, then you end up with the same list. We say that the set complement of a shape of class *a* has class 6-*a*. So, the gaps in the sponge are a cluster and the air between the blobs in a cluster is a sponge. It also means that the gaps between the branches in a tree create a tree structure, so the tree class is its own inverse in 2D.

In 3D, the list is just two classes longer: void, cluster, tree, sponge, *shell, foam* and solid. Here, cluster and tree have a similar meaning, and a sponge continues to

(1)		Void: contains region that is empty of points and is continuously scale symmetric	–
(2)		Cluster: a recursive, disconnected set of regions	convex
(3)		Tree: single connected structure with recursively branching limbs	convex and saddle
(4)		Sponge: single recursive network of connected limbs	saddle
(5)		Shell: single structure of recursively branching plates	saddle and concave
(6)		Foam: single recursive network of enclosing plates	concave
(7)		Solid: volume or area of space is filled, contains no finite sized gaps	–

Classification list for **3D scale-symmetric** structures: images, definitions and expected convexity.

be a scale-symmetric network of edges, so like a real sponge, it would be permeable to water. A shell is a scale-symmetric hierarchy of basins, so it is the set-wise complement of a tree. Finally, a foam is the structure that surrounds a cluster of 3D blobs, so it is not permeable to water, as each "air" pocket is isolated.

As with the 2D case, reversing the list is equivalent to taking the complement of the sets. This means that in 3D it is the sponge structure that is its own inverse.

While the given definitions of each class are descriptive rather than mathematical, they hopefully convey the structural character of each class. Their precise meaning can depend on what is being classified, for example, we can classify structures in nature if we allow the scale symmetry to be *approximate* and *local*. On the other extreme, for pure mathematical examples of each class, we can constrain the definition by the permitted curvature of the structure's surface, this has been listed in blue next to each class.

This system seems like a very nice and easy way to classify the structures in this book. They follow a simple progression of increasing self-connectivity, and the definitions are flexible enough to allow us to classify natural, as well as purely geometric examples. For example, the asteroid belt is well modelled as a *cluster*, while rivers, lightning and the tree structures of Chapter 3 are all in the *tree* class. The rings structure in Chapter 3, Figure 10, is a 2D *sponge*.

However, these lists are a very coarse categorisation, the five nouns — or the seven nouns in 3D — hardly allow the full variety of examples to be mapped out. On the other hand, a much longer list of classes would be hard to remember and would then fail to describe broad families. The way around this problem is to use a graded classification where the broad categories can be further split in order to better qualify each class.

This can be done by allowing each class name to be used as a qualifier in addition to its use as a noun. So, void-, cluster-, tree-, sponge- and solid- can now qualify the noun, and we can describe a structure as a void-sponge or solid-tree, for instance. This use of qualifier and noun expands the classification to a 5×5 table of classes for 2D objects and a 7×7 table for 3D objects. Objects can still be described broadly by their class as a noun or by their class as a qualifier. So, for example, we may describe an object as a *tree* (noun), or as *sponge*-like (qualifier), or specifically as a *sponge-tree* (qualifier-noun).

This reuse of a noun to qualify another noun is referred to as a *noun adjunct* and is common in everyday speech. For instance, *Her home is a guest house and his is a lighthouse but ours is just a "house-house," where we have house parties after doing the housework.* Here, the house noun is any sort of dwelling, but the house qualifier relates to a standard home. The conjunction of the two is sometimes used to clarify that you mean a standard house.

The same can be said of this classification. For a class *a-b*, the qualifier *a-* is defined as:

As an example in 2D, the cluster-cluster is a cluster structure and its complement is a sponge. This is a standard cluster like a 2D analogue of the asteroid belt. These classes *a-b* define a full classification table, and the cells on the downward diagonal give the standard examples of each class in the classification list.

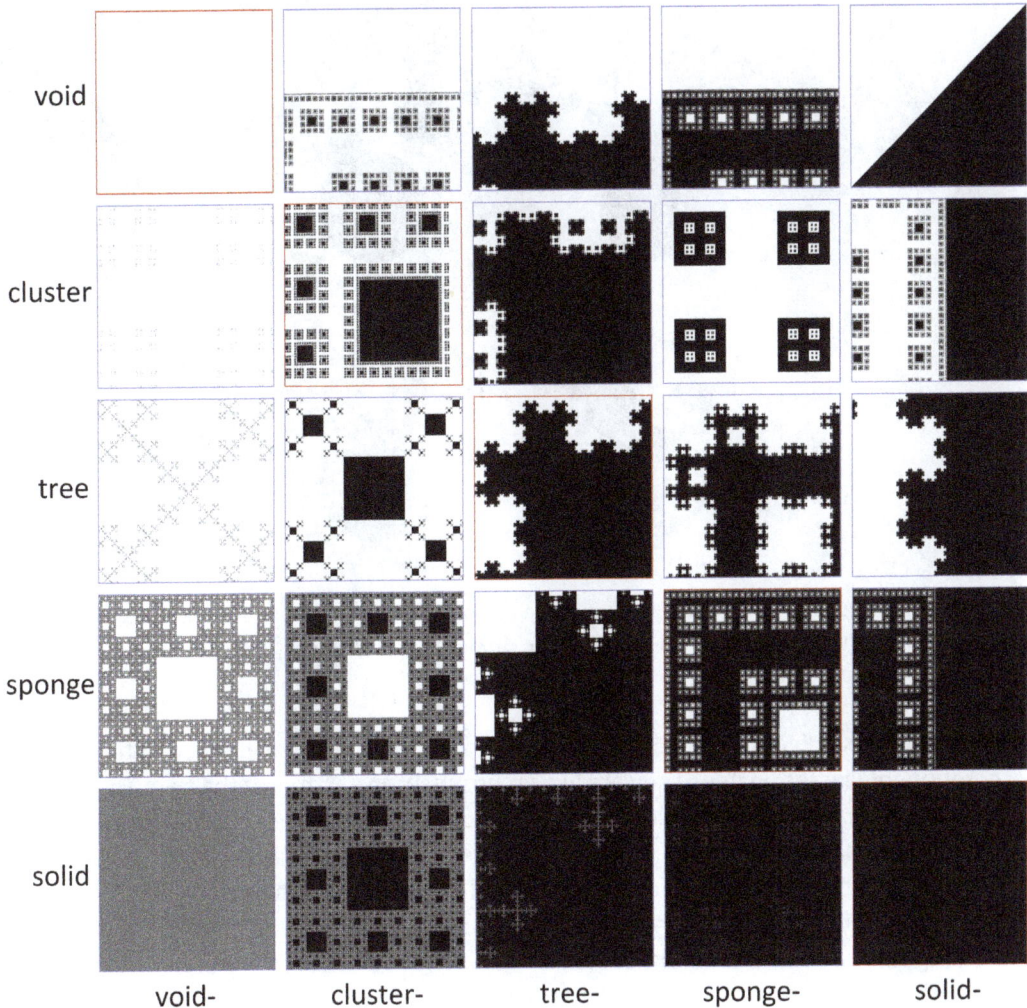

void

cluster

tree

sponge

solid

void-   cluster-   tree-   sponge-   solid-

Figure 1. 2D classification *table*, read as horizontal–vertical. Downward diagonal (red) are standard classes of the classification *list*. Images show some basic mathematical examples using recursion on a 3×3 (or larger) grid. The exact geometry is not important, only its connectivity.

Off the diagonal, we get more unusual variations. The cluster-tree, for example, is a tree structure, but its complement is a sponge. This is only possible when the pieces that make up the tree are connected at single points, an example of this is a tree of contacting disks, such as the Mandelbrot set. If the disk radii were any smaller, it would be a cluster-cluster, any larger and it would be a tree-tree. So, this class sits at a critical point between the two classes that describe it. All the classes below the downward diagonal have this property of being at a critical point between the two named classes.

Figure 2.   Example of below-diagonal and above-diagonal behaviour. The cluster-tree occurs at the critical point where a synchronised merge of the disks first contacts. The tree-cluster occurs on an unsynchronised merge, which equates to a cluster of trees.

Above the diagonal, the classes are not at a critical point between their two component classes, instead they are like an unsynchronised mix of the two. The tree-cluster by our definition is a cluster whose complement is a tree. It is a disconnected set of regions where the gaps contain recursive branches, so structurally speaking, it represents a cluster made from trees. Most of these above-diagonal classes a-b can be thought of as b made from a.

Together all these different cases form a 5×5 grid of classes in 2D, making a sort of periodic table of scale-symmetric structures. The table has a symmetry just like that of the classification list. In this case, around the upwards diagonal. This means that the set complement of a member of class a-b has class 6-b–6-a, the upwards diagonal is therefore the set of classes that are their own complement.

With this classification table, the seemingly surplus *void* and *solid* classes now become important, as you can see they are part of the definition of 16 of the 25 classes. The *void*-qualifier (the left column) describes the purely fractal version of each class, for instance, in Chapter 2, Figure 18, the tridendrite is a *void*-tree and the hexaflake is a *void*-sponge. By symmetry, the bottom row is the set complement of such fractals. Physically, it represents solids with differently structured cracks within, for example, the last image in Chapter 6, Figure 5, is a cluster-*solid*.

There is however the odd case of the void-solid class which is present in both lines. By the given definition, this means that the structure and its complement should both be solid. The only structures that fit into this definition are the *rational space-filling structures* from Chapter 3, an example being the set of rational 2D coordinates, there is also an example in the last image of Chapter 6, Figure 7. Such structures have no finite sized gaps and no dense regions with area and so satisfy the conditions of being a void-solid.

The top row cells are also unusual. The cluster-void, tree-void, and sponge-void have the properties of the void class without being empty. This occurs on the edge of a scale-symmetric shape. The empty half-plane satisfies the void definition given in the classification list, and the other half-plane satisfies the qualifier as long as the whole space remains scale-symmetric.

The right column is the set complement of this top row, so, for example, the solid-tree has a solid half-plane with a tree-like structure coming from it. We could call this a forest, with the half-plane representing the ground and the trees extending out of this, just as seen in Chapter 3, Figure 9.

We are again left with an unusual case at the conjunction of the two lines: the *solid-void*. The main example of this is a single solid half-plane.

Looking at the table as a whole, we can see that the downwards diagonal shows a progression from empty space to filled space, and the upwards diagonal shows a progression from a homogenous mixture towards a complete separation of the two colours.

You may have noticed that fractal curves like the Koch curve are missing from this table. Curves in general are not scale-symmetric shapes if they have thickness. They can be made scale-symmetric if they taper, but then they have all the properties of being a tree with only one branch per junction. By extension, a thin fractal curve like the Koch curve would be classified as a special type of void-tree. Fractal curves also feature as the boundary of the structures in the table, for example, the 1.5D curve in Chapter 2 Figure 15. is the boundary of a tree-tree.

A similar table can be generated for three-dimensional objects, the addition of the shell and foam classes extends the size to a 7×7 table.

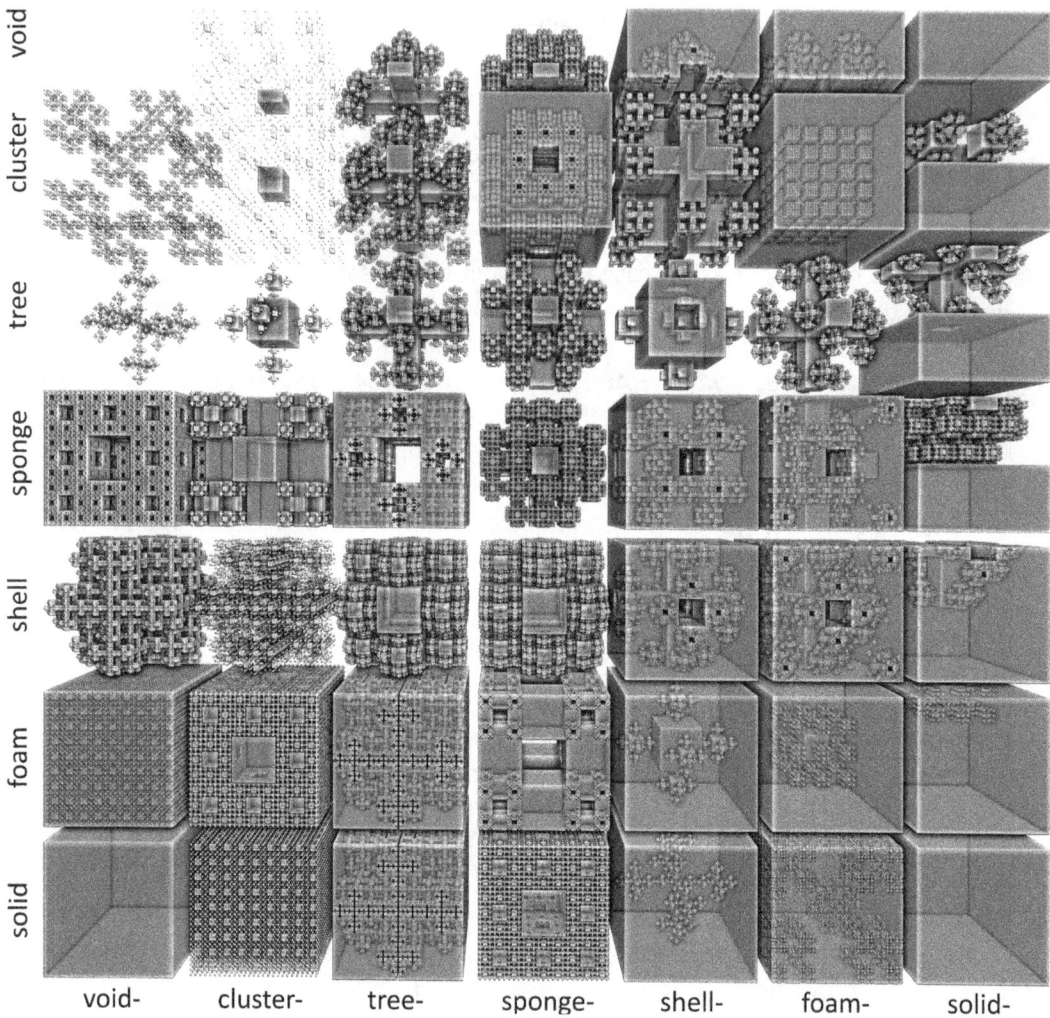

Figure 3.   3D structure classification table using cubic substitution rules as examples. Rendered with transparency to visualise the inner structure.

I won't detail the different examples in the above image as there are too many, and the specific shapes aren't important, nevertheless you may see that most of these are created through a recursive rule on 3×3×3 voxels or 5×5×5 voxels. Looking back at some examples in the book: in Chapter 4, the bubbles in Figure 14 are a void-foam, the Kleinian shapes in Figure 15 are void-sponges, and the sphere tree in Figure 17 is a type of sponge-tree. The first and third rocks in Chapter 6, Figure 19, are a sponge-sponge and a shell-tree, respectively. Shell-trees are a common class, they include shapes with hierarchies of both protrusions and indents, so they represent bumpy rocks well.

★

With these tables established, let's look at where scale-symmetric structures that have been studied fit in. In the following tables, I have included examples from this book as well as existing well-known and named structures. The latter set are based on shapes known to me and also by conducting a survey with the online community at *fractalforums*,[1] so it is not a rigorous set, but it is nonetheless indicative of the landscape of studied structures. The names of all of these are available at the end of the chapter.

Figure 4. The classification of the 2D structures studied in this book and elsewhere.

[1] fractalforums.org.

It is noticeable that the left-hand column is well represented, as these are the fractal structures, which are usually simpler constructions than the non-fractals. Of the other areas, trees are most common, as seen in the central row and in the central column. Surprisingly, even for the basic cluster-cluster and sponge-sponge it is hard to find a named recursive structure that represents these classes.

In 3D, the set of examples is sparser, most likely because three-dimensional structures are generally harder to visualise, manipulate and reason about.

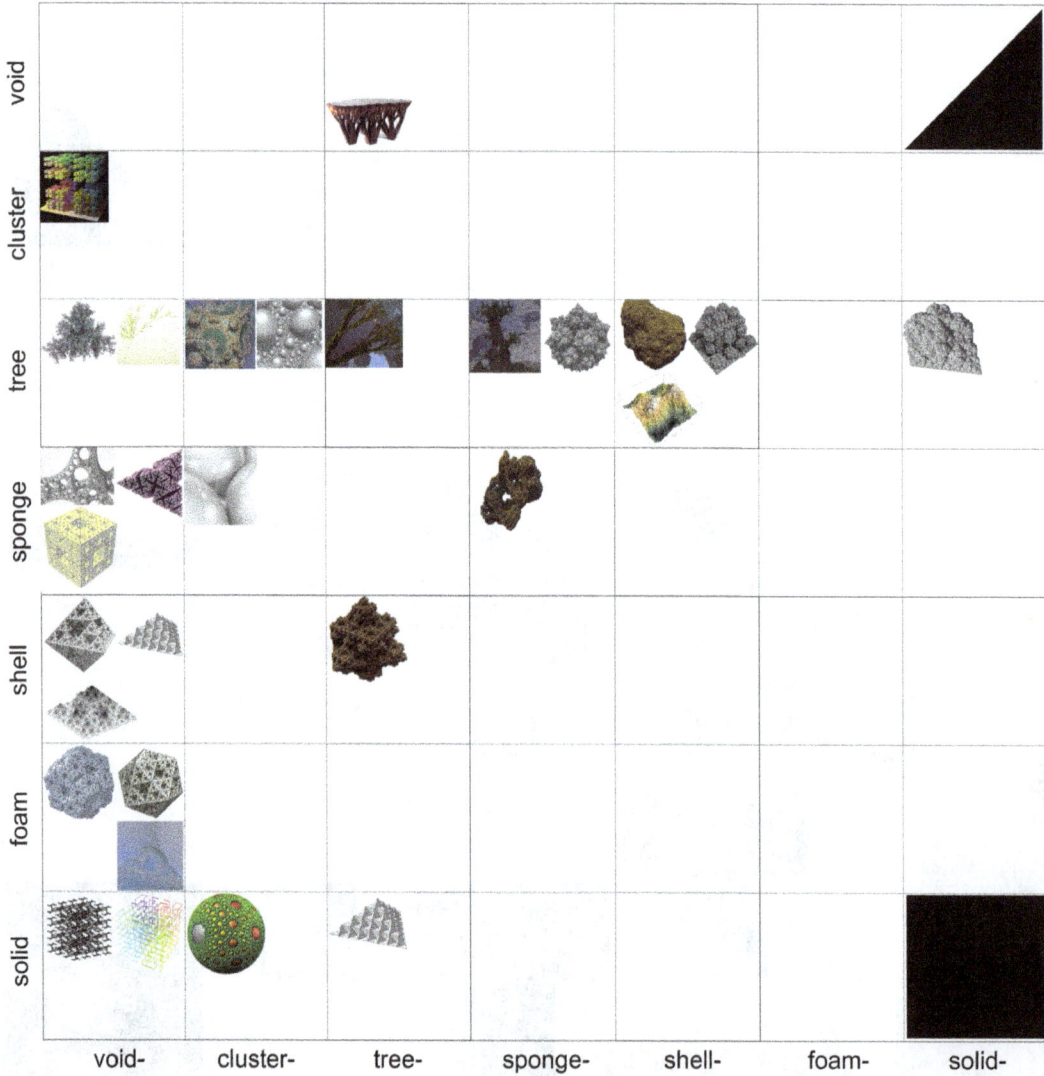

Figure 5.   Classification of the 3D structures studied in this book and elsewhere.

Again, the left column is well represented, and there are many examples around the tree row and column. But much of the lower-right half of the table is lacking examples. This may reflect the difficulty in visualising structures that are more solid, and so details are obscured. This shouldn't matter from a research point of view because one can typically study the set complement instead. Nevertheless, improved visualisation tools could help the research in 3D, for example, the use of transparency, refraction and Fresnel-like reflection models can help to understand and present the shape of the denser classes of structure, just as it helped with the bubbles structure in Chapter 4.

Since so many scale-symmetric structures are found in nature, it seems reasonable that most of these classes could also be found naturally. If we allow approximate and local scale symmetry, then this does seem to be the case. For instance, (galaxy) clusters, trees, sponges, shells, foams, and solids all occur in nature. The shell here is not a typical sea shell, but the shell surrounding any sort of tree structure. A walnut shell is a reasonable example, it has the basins-within-basins structure that characterises the class.

Off-diagonals occur too, in 2D a forest is a tree-cluster from above, and a solid-tree when viewed from the side, with the ground included. The classes below the downwards diagonal are often represented through contacts, for instance, the cluster of connected stones on a scree slope forms a cluster-sponge.

So, how could we summarise these rather detailed looking tables? One way is to maintain a metaphor for the four corners. We can think of the void-void as air, the solid-solid as rock and the solid-void as land — the place where solid and air meet. I also like to think of the void-solid as a bit like water, it lacks the dense connectivity to keep it rigid, but also has no air gaps to make it compressible. This analogy is aided by its proximity to froth and bubbles, which fit into the adjacent class: void-foam.

This metaphor makes the table a wonderfully neat representation of much of nature: The table is framed by "air," "land," "water," "rock," and the other classes are intermediates between these four. More abstractly, we might just think of these four corners as representing sparsity, heterogeneity, homogeneity and density, respectively.

★

So far, we have neglected the topic of moving structures, such as those in Chapters 6 and 7. However, since the exact geometry of each class is not constrained, it seems reasonable that a moving structure can be a member of the above tables as long as its inter-connections don't change in the process. For example, we allow clusters to move as long as they don't collide and we allow trees to sway as long as the branches don't hit together. We still require the structures to be scale-symmetric though, and time must be included in the scale symmetry. So, for swaying trees, the small twigs should sway with a proportionally lower time period than the larger branches, just as in the tree example of Chapter 7.

Those structures that *do* alter their connectivity could be described by the set of classes that they occupy over their life span. However, a more systematic approach is to introduce time into the structure as an extra dimension. For instance, animating 2D structures are considered as 2+1D, where the +1 here refers to the single time dimension. The extra time dimension acts much like adding an extra spatial dimension, so in this case, the 2+1D structures are almost the same set as the 3D structures, and we use the same class labels. However, we constrain the velocity of the structure to be everywhere finite, or more realistically, less than the speed of light. For brevity, let us call these scale-symmetric animating structures *behaviours*.

The addition of the time dimension means that such 2+1D behaviours are classified in a 7×7 table and 3+1D behaviours occupy a 9×9 table. That is a lot of classes, so let us have a look at just the classification *lists*, of which there are just five classes in 1+1D:

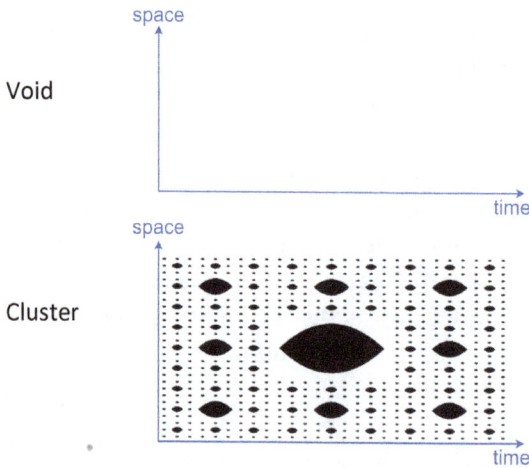

Figure 6.   Classification list for 1+1D scale-symmetric structures. These are 1D structures that move.

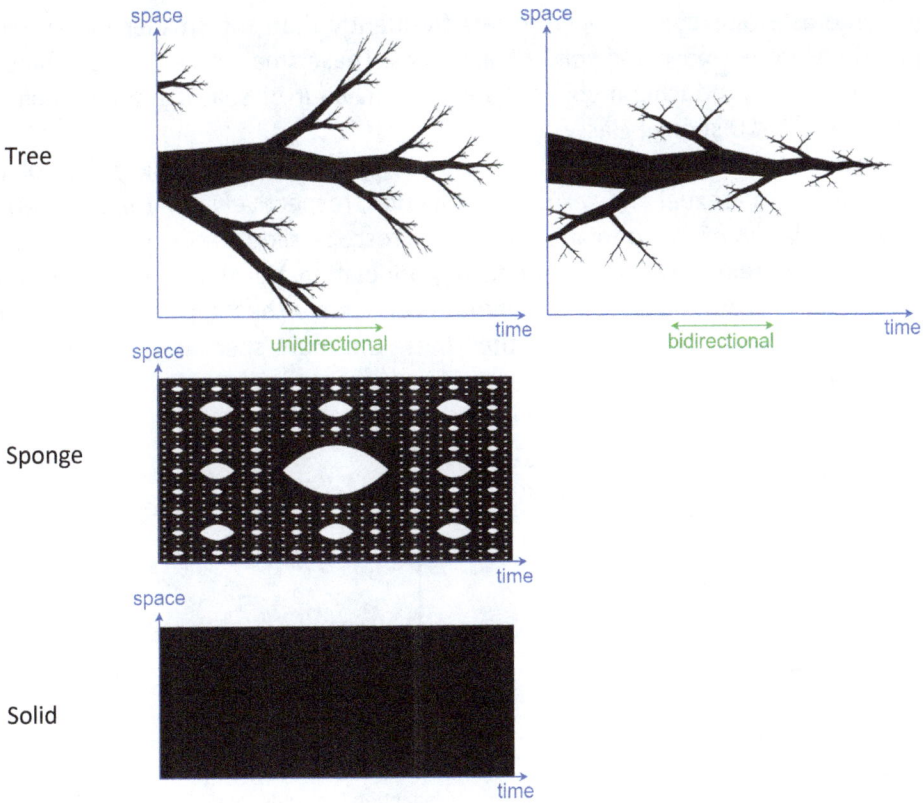

Figure 6.   (*Continued*)

The above behaviours in 1+1D act similarly in higher spatial dimensions. There are three non-trivial behaviours — the cluster, tree and sponge — let's look at each one in turn:

The *cluster* is a distribution of blobs that appear, grow and shrink independently. In 1+1D, these blobs are just line segments, but in 2+1D, they are regions, they act rather like the emergence and evaporation of puddles, ponds and lakes in a landscape. There are usually many more small puddles than large lakes, and the puddles come and go much more frequently than the lakes.

The *tree* is a distribution of splitting or combining shapes, but never recombining shapes. These shapes are mainly blobs, but they could also be tree shapes. The blob-like automaton in Chapter 6, Figure 24 (middle), is an example of a 2+1D tree, in this case, the blobs are combining together over time.

The *sponge* appears as a set of splitting and recombining shapes, these shapes could be blobs or trees or even sponges, though they are of course only line segments in 1+1D. The lava-like automaton in Chapter 6, Figure 24, is an example of a 2+1D sponge. An example in 3+1D is the asteroid belt in our solar system. In this belt, large asteroids collide together less frequently than the smaller meteoroids, and fast collisions smash and split the asteroids. These smaller pieces recombine in slower impacts, producing a stable dynamic behaviour of splitting and combining that is in the 3+1D sponge class.

These last two behaviours — *tree* and *sponge* — have an analogy in biology, being rather like asexual and sexual reproduction, respectively. An asexually reproducing population of microorganisms on a microscope slide traces out a 2+1D tree over time, whereas a sexually reproducing population traces out a 2+1D sponge since organisms must recombine to mate, even if they then separate again. If we disregard the size of the organisms, then these are more specifically *void-tree* and *void-sponge* behaviours.

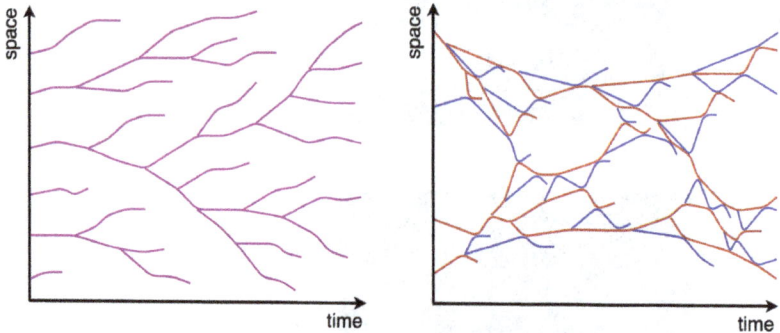

Figure 7.   Asexual reproduction (left) and sexual reproduction (right) depicted here as 1+1D void-tree and void-sponge behaviours. The blue and red trajectories are male and female, respectively.

In 2+1D there are two further behaviours: *shell* and *foam*. These are just the set complement of the *tree* and *cluster* behaviours, respectively. Similarly, the three denser 3+1D behaviours are just the set complement of the 3+1D *sponge*, *tree* and *cluster* behaviours. However, there are seven non-trivial behaviours in the 3+1D classification list, so there remains an interesting final case, the 3+1D shell.

This *shell* appears like a 3D sponge, where the limbs of the sponge repeatedly flatten and split apart, creating a new hole. It may also do the opposite and combine limbs together, but it never recombines them. Depending on the thickness of the limbs, it may have areas that are shell-like or tree-like, but it will always contain loops and is always a single connected structure.

Since the *cluster*, *tree* and *sponge* behaviours mirror some important behavioural patterns in nature, it is interesting to think about whether this *shell* behaviour also has a natural counterpart. Since biological behaviours on land are somewhat two-dimensional, we might look to the sea instead for examples. Sea sponges are a good example. They have a sponge structure that develops new holes as they grow. We know this because sponges often grow from a broken fragment of their parent, and if they didn't develop new holes, then the children would eventually lose all their holes. Sea sponges are therefore 3+1D *shells*.

Figure 8.   Deep sea sponges grow and develop new holes, which is a 3+1D *shell* behaviour.

Another candidate for a 3+1D shell is the large-scale structure of the universe. It is not a uniform distribution of galaxies but is instead like a giant sponge over a wide range of scales from about 20 light years upwards. It is built of long *galaxy filaments* of millions of galaxies and some flat *galaxy walls,* which can split and form new filaments. These are the sort of structures we would expect from a 3+1D shell behaviour.

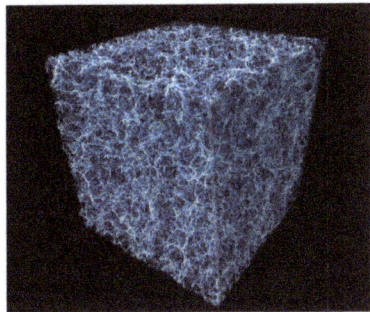

This *shell* completes the list of behaviour classes up to 3+1D. Note that these simple *tree*, *sponge* and *shell* classes for moving structures appear more complex at any time instant. The 3+1D sponge, for example, may look like any mixture of blobs, *trees,* and *sponge* structures. A similar effect was noted in Chapter 7, where simple symmetries in swaying trees and rippling ponds were not evident at any single time instant. The implication of this effect is that more structures in nature may follow a simple scale-symmetric design than we currently notice. It tells us that when characterising natural features like clouds, mountains, or forests, we ought to look at their long-term behaviour over time and not just their current shape.

★

We now have a quite comprehensive method of classifying static and moving scale-symmetric structures, which is broad enough to capture most of the structures in this book. Next, we will look into how to measure and quantify all these structures, and so allow us to summarise their most important characteristics.

Appendix: Table References

These listings provide details on where each shape is found or the name of established shapes. *Italicised* names can be found in the glossary of shapes.

5×5 Table of 2D Structures

solid-void:	A half-plane.
void-cluster:	Chapter 3, leaves of tree Chapter 10, Mobius set Chapter 5, 2D *Cantor set.*
tree-cluster:	Quasi-octagonal automaton Chapter 6.
sponge-cluster:	Square automaton Chapter 6.
void-tree:	Chapter 9, Tridendrite Chapter 2, binary tree Chapter 3, *Vicsek fractal.*
cluster-tree:	*Mandelbrot set.*
tree-tree:	Cantor-tree Chapter 3, *Koch snowflake*, Chapter 6, Cesàro fractal.
sponge-tree:	Quasi-octagonal automaton Chapter 6, disk-tree Chapter 4.
solid-tree:	Forest structure Chapter 3.
void-sponge:	Chapter 6, *Menger carpet, Sierpinski triangle*, bubbles Chapter 4.
cluster-sponge:	Chapter 4, Chapter 3, *Ford disks.*
tree-sponge:	Quadric cross.
sponge-sponge:	Harris spiral, its limit self-contacts to form a sponge.
void-solid:	*Hilbert curve, Peano curve,* Chapter 6, Pinwheel tiling.
cluster-solid:	Square automaton Chapter 6.
tree-solid:	Inverted *Vicsek fractal*, T-square, automaton Chapter 6.
sponge-solid:	Inverted 2D *Cantor set.*
solid-solid:	The filled plane.

7×7 Table of 3D Structures

tree-void:	This fractal table is a tree below and void above the surface.
solid-void:	A half-volume.
void-cluster:	3D *Cantor dust.*
void-tree:	*Diffusion Limited Aggregation*, L-system tree.
cluster-tree:	Scale −1.5 Mandelbox corner Chapter 5, $k < 1$ *Ford spheres* Chapter 4.
tree-tree:	Within scale −1.5 Mandelbox Chapter 5.
sponge-tree:	Within scale −1.5 Mandelbox Chapter 5. Sphere tree Chapter 4.
shell-tree:	Chapter 6, 2.5D surface (filled) Chapter 2, fractal landscape volume.
solid-tree:	Extended *pyramidal surface* Chapter 2 (filled inside and below).
void-sponge:	Bubble edges Chapter 4, Delta fractal, *Menger sponge.*

cluster-sponge:	*Ford spheres.*
sponge-sponge:	3D automaton Chapter 6.
void-shell:	Octahedron fractal, Koch surface fractal, quadratic surface.
tree-shell:	Filled quadratic surface.
void-foam:	Dodecahedron fractal, Icosahedron fractal, bubbles Chapter 4.
void-solid:	3D H fractal, 3D Moore curve.
cluster-solid:	Apollonian (filled) sphere packing.
tree-solid:	Filled Koch surface.
solid-solid:	Filled volume.

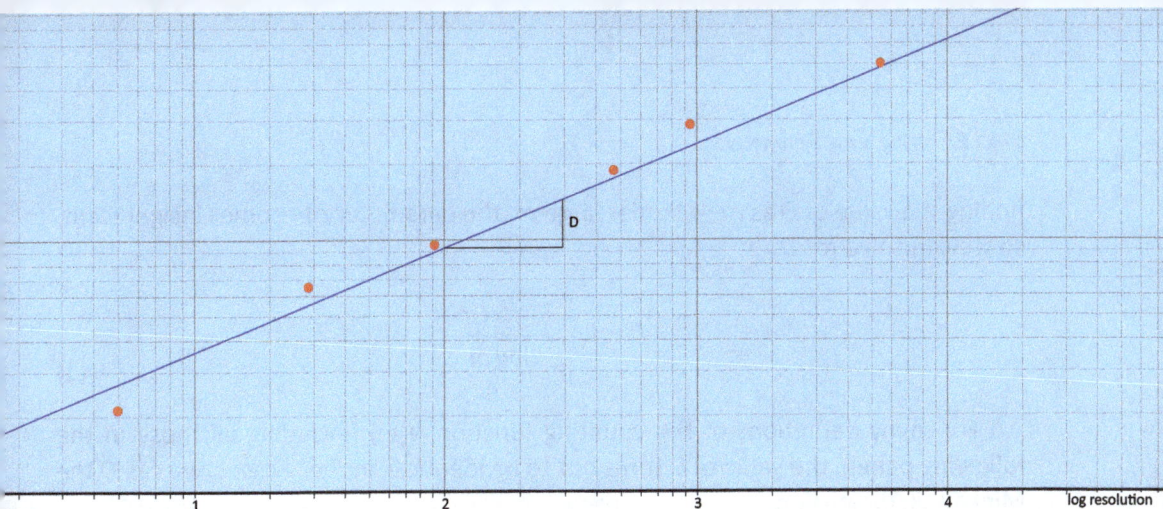

Chapter 9

Quantification

Can geometry deliver what the Greek root of its name [geo-] seemed to promise—
truthful measurement, not only of cultivated fields along the Nile River but also of
untamed Earth?

— Benoit Mandelbrot

This chapter is about quantifying scale-symmetric structures. Not only the mathematical ones in this book but also the ones found in nature, such as mountains, rivers and lightning. The aim is to describe these structures through a set of continuous attributes in order for us to measure, organise and compare structures. It therefore has a complementary role to the discrete classifications of Chapter 8.

The general approach to quantifying self-similar structures is to find a function that models their growth rate and then quantify each structure by the parameters that best fit the model. For fractals, the model is a simple power function:

$$N(m) = cm^D$$

In this model, D is the *fractal dimension* of the fractal and c is its *fractal content*, which tells us how much fractal there is. $N(m)$ is the counting function of the set in question for a given resolution m, it may be the number of pixels of width m^{-1} that cover the set or the number of covering balls of radius m^{-1}. The precise meaning of c and D depend on the choice of $N(m)$.

For non-trivial fractals, the count does not neatly fit this power law, so we must find the best fit parameters. The best fit D over all resolutions m is equivalent to finding the value of D as $m \to \infty$. For large m, the parameter c becomes insignificant, so we can solve for D:

$$D = \lim_{m \to \infty} \frac{\log N(m)}{\log m}$$

(1)

For many definitions of the counting function $N(m)$ (including all those in the following panel), the value of D turns out to be identical, in these cases, we call D the Minkowski dimension.

Now that D is calculated, the fractal content is found by substitution:

$$c = \lim_{m \to \infty} \frac{N(m)}{m^D}$$

(2)

$N_{\text{box}}(m)$ **the box-counting number** is the number of pixels of width m^{-1} that cover the set.

$N_{\text{ext}}(m)$ **the external covering number** is the fewest number of balls of radius m^{-1} that cover the set.

$N_{\text{pack}}(m)$ **the packing number** is the largest number of disjoint balls radius m^{-1} with centres in the set.

$N_M(m)$ **the Minkowski number** in R^n is m^n times the n-dimensional volume of all points within radius m^{-1} of the set.

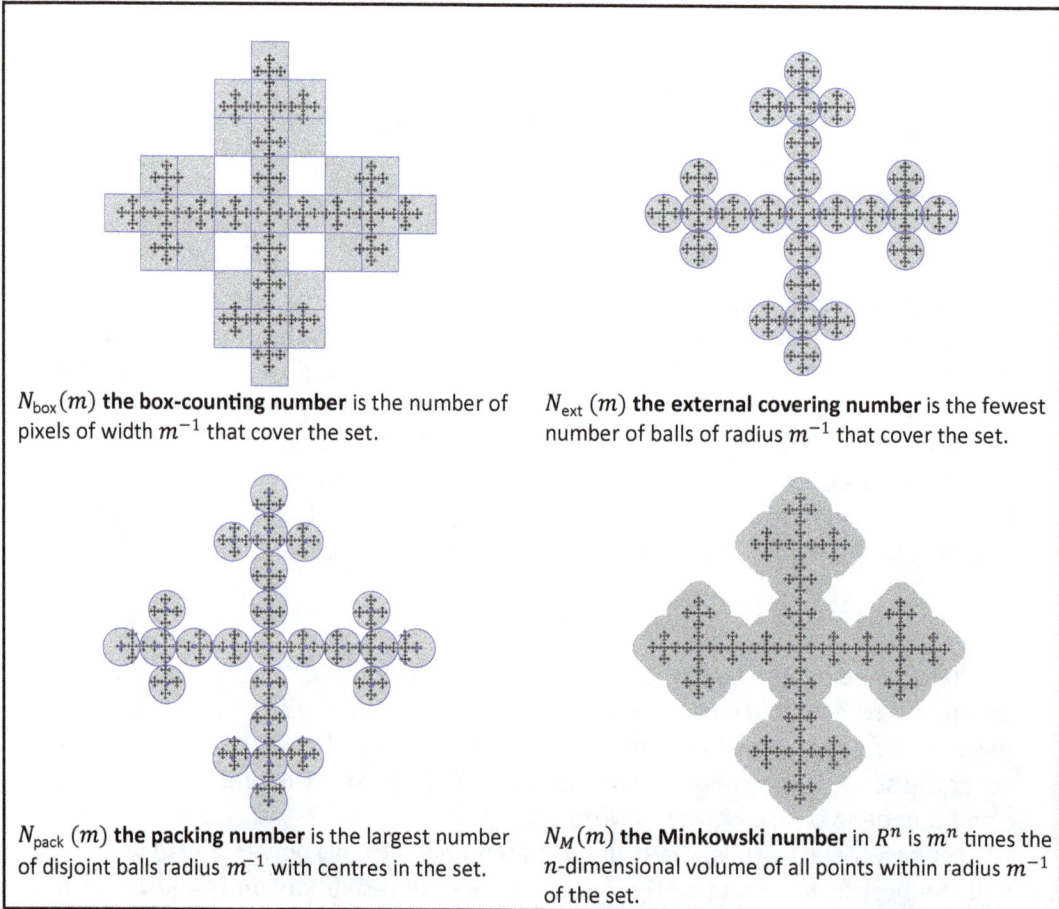

Panel 1. Some different types of counting functions, these can all be used for the Minkowski dimension.

Fractal content extends the notions of count, length, area and volume to non-integer dimensions, giving a general notion of the amount of structure. Unlike the dimension, the fractal content depends on the choice of counting function. The clearest and most well-studied choice of $N(m)$ is the *Minkowski number* $N_M(m)$, its value is m^n times the n-dimensional volume of the points within radius m^{-1} of the set. So, in 2D, this is m^2 times the area of the set expanded by m^{-1}. This choice defines c as the *Minkowski content*.

Normalised Minkowski Content

To properly generalise the notions of length, area and volume, it is necessary to scale the fractal content by a constant value, so that lines, disks and balls have their usual size. In the Minkowski case, this *normalised Minkowski content M* is achieved by dividing by the volume α of an $n - D$ dimensional ball of unit radius:

$$\alpha = \frac{\pi^{\frac{n-D}{2}}}{\Gamma\left(\frac{n-D}{2}+1\right)}$$

where $\Gamma()$ is the Gamma function. So, $\hat{c} = c/\alpha$.

With this normalisation, we can use $\hat{c}$ to describe the *size* of the set rather than a counting function of the set:

$$size = \hat{c}m^D$$

where *size* can be the length, area or volume, depending on the topological dimension of the set.

Let us now apply these formulae to the simple case of *substitution rule* fractals, such as the first examples in Chapter 1. These fractals are generated by replacing a shape A by n smaller copies, scaled down by a factor of s.

Figure 1.   *Substitution rule* generating a fractal limit set, iterations 0, 1, 2 and ∞.

From the rule description, we can see that at high enough resolutions ($m \to \infty$);

$$N(sm) = nN(m)$$

This recurrence equation is solved as $N(m) \propto m^{\log_s n}$. Applying Eq. (1) then gives us $D = \log_s n$, which is the *similarity dimension* that was used in Chapter 2.

To find the *Minkowski content*, we consider the ratio $\frac{N_M(m)}{m^D}$ with respect to log m:

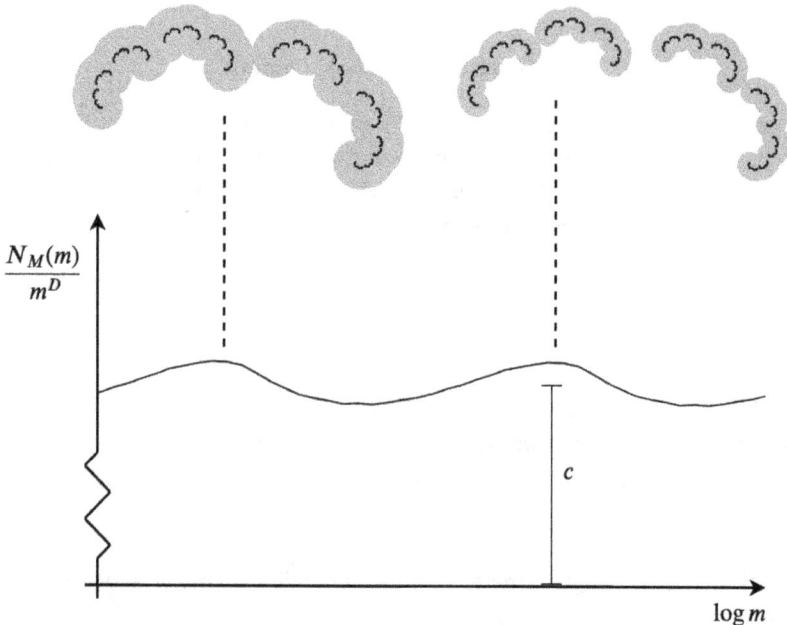

Figure 2.   Graph of the local *Minkowski content* at each resolution m, showing the grey volume that the Minkowski number measures at two peaks. The mean height c is the *average Minkowski content*.

As you can see, the ratio does not converge with increasing log m, instead it oscillates. This means that the limit does not exist in Eq. (2) and so we cannot use it as a simple means to find the best fit c. Instead, we have to take a numerical average over the whole domain $0 \leq \log m < \infty$ to obtain the *average fractal content*. The oscillation peaks sit at the critical radii where parts of the expanded volumes are "kissing," this event repeats for every scaling by s, giving the oscillatory pattern.

So, even for the simple *substitution rule* fractals, the model $N(m) = cm^D$ does not tell the whole story. For extra descriptive power, an extended growth function can be used, there are two such extensions.

The first is to model these observed oscillations of the Minkowski content, primarily by a sine wave of best fit, but more generally by a sum of sine waves. This model is equivalent to the sum of a set of power functions with complex exponents, and so this is the research topic of *complex fractal dimensions*.

I recommend the work of Lapidus [28] and Van Frankenhuijsen [29] for an in-depth discussion of the topic. These complex dimensions reveal insights about the scaling properties of the set, but from a modelling point of view, the oscillation is often negligible, in the above case, it only represents a variation of 0.4%.

The second extension is to model secondary characteristics of the growth function beyond the single power function. Consider the 2D case of a filled unit square, the expanded area is $1 + 4r + \pi r^2$, so its Minkowski number grows as $N_M(m) = 1m^2 + 4m + \pi$. The first term here gives the dimension and content, the other terms represent secondary 'curvatures' of the square. These are multiples of the *intrinsic curvatures* of the set, which are a well-studied set of characteristics. In this case, they are its area, perimeter and *Euler number* in 2D.

We can capture these characteristics using variants of the Minkowski number $N_M(m)$. If X_M is the set of points within distance m^{-1} of set X, then:

$N_{M2}(m)$ is the area of X_M, times m^2,
$N_{M1}(m)$ is the length of the perimeter of X_M, times m,
$N_{M0}(m)$ is the *Euler number* of X_M: the number of separate sets minus the number of holes.

Calculating c and D using these counting functions gives the *fractal curvatures* and their *fractal scalings*, respectively. Here, $N_{Mn}(m) \equiv N_M(m)$ for sets in $\mathbb{R}^n$, so the primary *fractal curvature* is the *fractal content*.

This parameterisation of fractal sets is expounded upon by Peter Straka in the paper "Fractal Curvature Measures and Image Analysis" [30], it can be quite revealing, for instance, parameters associated with N_{M0} appear to be well suited to distinguishing between the cluster, tree and sponge classes in the previous chapter's 2D *classification list*. Nevertheless, these additional counting functions are harder to calculate than just N_M and are usually less accurate.

Fractal Content or Fractal Measure?

When the fractal content of a set satisfies the properties of a mathematical *measure*, we can refer to it as the set's fractal measure. The Hausdorff measure is the main example of this, however this measure is very difficult to calculate, even some simple fractals only have an estimated range, so in this book, we use the simpler Minkowski content.

Following are some of the fractals with known Hausdorff measure [31–33]: the Cantor middle-third set, the Koch curve, the 2D Cantor middle-half set and the Sierpinski triangle, all of width m. Figures to 2.d.p.

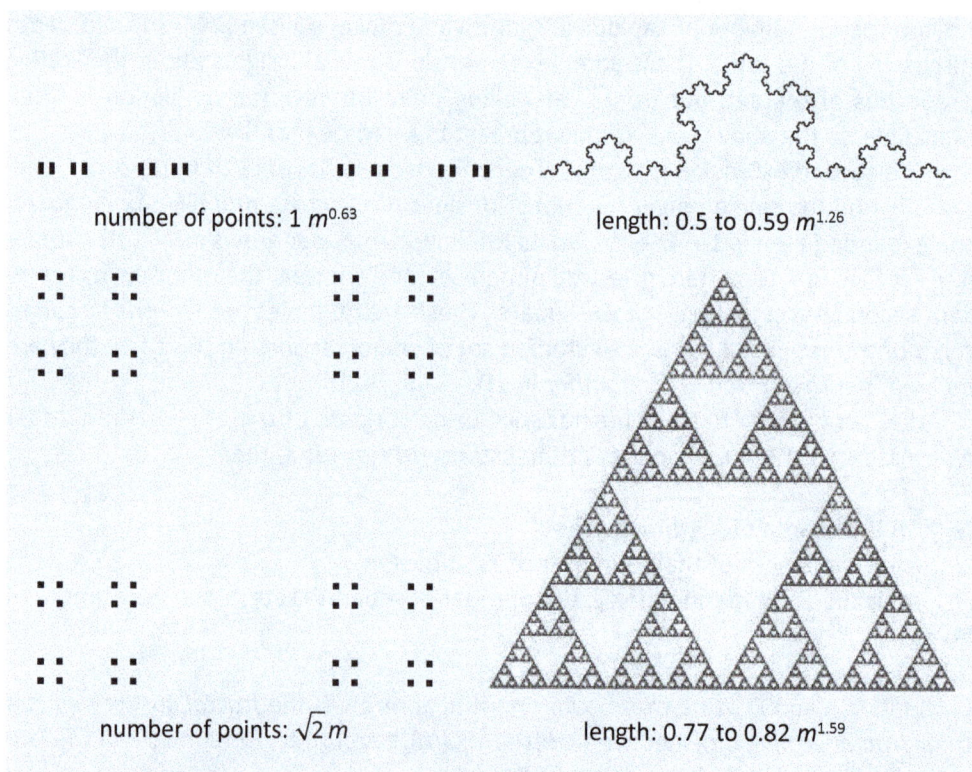

number of points: 1 $m^{0.63}$ length: 0.5 to 0.59 $m^{1.26}$

number of points: $\sqrt{2}\,m$ length: 0.77 to 0.82 $m^{1.59}$

Even without these two extended models, we are forced to consider new growth models when quantifying *non-fractal* structures, as the growth of these structures doesn't follow a simple power law:

Let's look at the example of the *multiplication rule* construction technique, which is the basic method behind the recursive structures of Chapter 3. This construction starts with a base shape A of fractal content a and dimension d, then iteratively substitutes the set with n copies scaled down by a factor of s, while keeping a copy of the base shape in its original location:

$$X_{i+1} = A \cup \bigcup_{j=1}^{n} s^{-1}T_j X_i$$

for Euclidean transformations T_j.

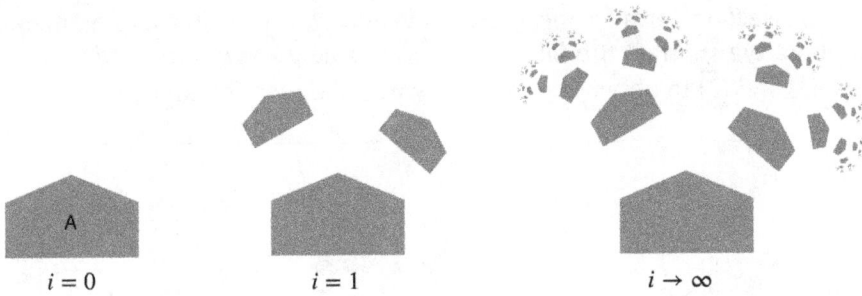

Figure 3. *Multiplication rule* example with a 2D base shape, so $d = 2$, and here, $n = 2$ and $s = 2$.

For sufficiently high resolution ($m \rightarrow \infty$), the structure is defined by the recursion:

$$N(m) = nN\left(\frac{m}{s}\right) + am^d$$

which has three solution cases:

$$s^d > n: \; N(m) = c_0 m^d - k_0 m^{\log_s n}$$
$$s^d = n: \; N(m) = c_1 m^d \log m + k_1 m^d$$
$$s^d < n: \; N(m) = k_2 m^{\log_s n} - c_0 m^d$$

where $c_0 = as^d/|s^d - n|$ and $c_1 = a/\log s$. The best-fit constants k_i are solved for in the same manner as Eq. (2).

In each case, the growth is not a simple power function but is instead composed of two terms. The first is the dominant growth term and the second term (in grey) provides the parameters to further quantify the structure, if we so wish.

This solution works when you already know the construction variables a, s, d and n, but when they are not known, or when we are looking at a more general scale-symmetric shape, then we need more general formulae, akin to those in Eqs. (1) and (2).

In fact, Eqs. (1) and (2) correctly find the exponent and coefficient of the dominant term, except in the $s^d \neq n$ case. Here, Eq. (2) yields a fractal content of infinity, which prevents such shapes from being measured or compared. So, in this case, we need to use a modified equation to extract the coefficient:

$$c = \lim_{m \to \infty} \frac{N(m)}{m^D \log m}$$

For the second term in the growth function, we can find its parameters by substitution. We subtract the first term (which is now known) from $N(m)$ and then use Eqs. (1) and (2) to obtain the secondary fractal dimension and content.

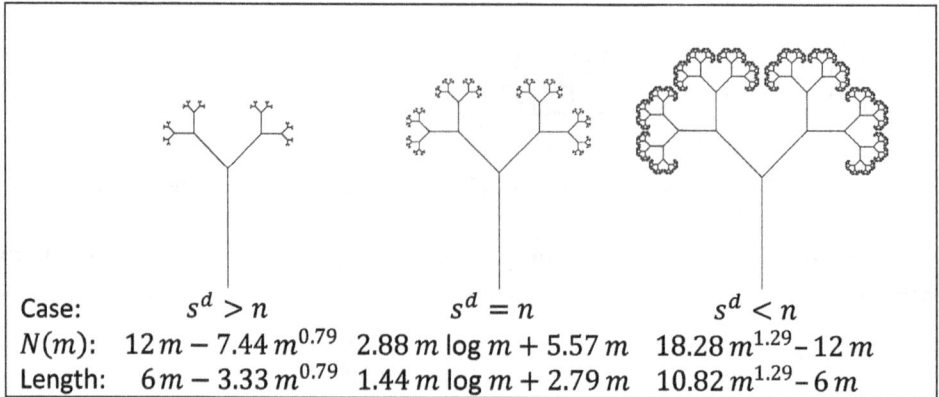

Case:	$s^d > n$	$s^d = n$	$s^d < n$
$N(m)$:	$12\,m - 7.44\,m^{0.79}$	$2.88\,m \log m + 5.57\,m$	$18.28\,m^{1.29} - 12\,m$
Length:	$6\,m - 3.33\,m^{0.79}$	$1.44\,m \log m + 2.79\,m$	$10.82\,m^{1.29} - 6\,m$

Figure 4.   Approximate growth law for binary trees of unit trunk length, with $a = 2$, $d = 1$, $n = 2$, for the cases of $s = 12/5$, $12/6$ and $12/7$ from left to right. Top: un-normalised, bottom: normalised fractal contents.

We now have three different types of secondary growth characteristics for non-fractals: the oscillations described by complex dimensions, the *fractal curvatures*, and the secondary exponents just discussed. All three are quite complex to calculate and to interpret, so I suggest a fourth, simpler approach.

The fourth approach is to recall from Chapter 3 that every non-fractal scale-symmetric set has one or more subsets that are fractal. So, we can quantify these shapes by extracting each subset as its own set, and extracting its fractal dimension and fractal content using Eqs. (1) and (2). For tree structures, these subsets can include the boundary and the leaf points. We could also quantify the total branch length of the tree or the number of branch points.

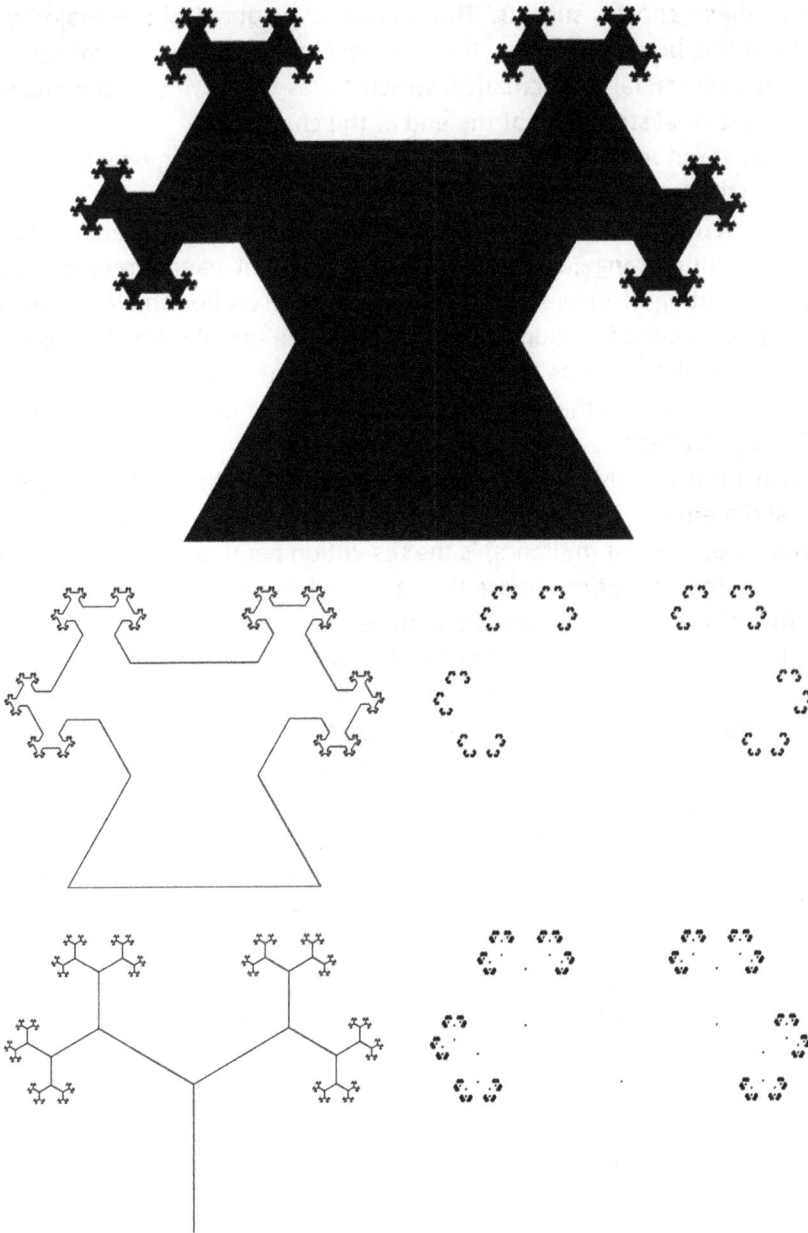

Figure 5.   Triangle binary tree and four subsets that that measure, in order: the perimeter, the number of leaves, the branch length and the number of branch points.

This is a flexible approach that allows us to quantify an attribute of a structure by representing it as a set. So, apart from the above $s^d = n$ edge case, most scale-symmetric structures can be quantified by the fractal dimension and content of the set and of these chosen subsets. This allows us to quantify the majority of the structures in this book and allows the well-developed tools of fractal geometry to apply to the non-fractal mathematical structures as well. We will demonstrate this approach on several structures at the end of the chapter.

But what about non-mathematical structures? In particular, how do we quantify the physical structures in nature?

These structures show *approximate, local scale symmetry*, where the symmetry exists over a limited range, so we must find the best fit parameters to the growth model by computing $N(m)$ over the limited range of resolutions m. When the growth model is a single power function $N(m) = cm^D$, then the samples $(m, N(m))$ can be plotted in a log–log plot, such as the one shown above the chapter title. The gradient and y-axis intersection of the line of best fit can then be used to estimate the fractal dimension and content.

You may have already noticed the similarity between the model cm^D and the use of physical dimensional quantities in real-world measurements such as a 4 m^2 carpet. In this case, the first m (italics) is the resolution parameter and the second m is the unit of metres m, but otherwise they are used in a very similar way.

Traditionally, these dimensional quantities have only applied to integer dimensions, but equating these quantities with the normalised fractal content and fractal dimension of the objects has been in practice for quite some time. For example, in 1983, G. I. Barenblatt used fractal quantities to measure the specific absorbing capacity of the respiratory organs of certain marine life by modelling these organs as a fractal. In 1992, F. M. Borodich measured the surface energy of cracks, modelling the crack as a fractal and describing the crack's surface *area* using its fractal content and dimension. Since then, fractal parameters continue to be used in the physical sciences, but the approach is complicated by a lack of consensus on which type of fractal content and dimension is appropriate.

However, if we make it clear which type we are using, then this representation as a dimensional quantity makes for a succinct and natural means of quantifying real-world scale-symmetric structures. For example, using the *yardstick method*, the *length* of the western coastline of Great Britain can be measured as roughly 5000 km$^{1.25}$. This is equivalent to representing the coastline by a fractal curve of dimension 1.25 with a normalised fractal content of 5000.

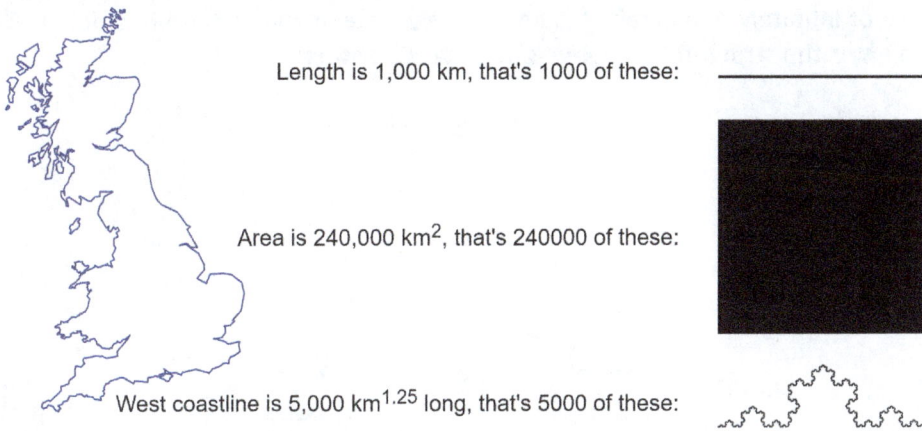

Length is 1,000 km, that's 1000 of these: —————————

Area is 240,000 km² , that's 240000 of these:

West coastline is 5,000 km$^{1.25}$ long, that's 5000 of these:

Figure 6.   Measuring Great Britain using the yardstick method for the fractal dimension and content.

So, the quantification of scale-symmetric structures is ultimately described in the same way for mathematical and natural structures: as the best fit parameters to a power function. This equivalence raises an interesting question however.

> If the exponents of dimensional quantities equate to the fractal dimension of the representative structure, then what sort of structure is the one that is measured in m^{-1}?

After all, we measure in m^{-1} all the time in the sciences, such as when measuring the number of leaves per metre along a vine. Furthermore, what sort of structure is being measured in negative *fractional* dimensions? For instance, if we approximate the coastline of Britain as having fractal dimension 1.25, then the number of fence posts per shore length should be measured in m$^{-1.25}$. There is no shape that produces negative dimensions with the methods presented so far, so a new dimension formula would be required.

For a complete picture of scale-symmetric structures, we should really consider structures represented with dimensional quantities of any exponent, not just the positive ones. So, in the spirit of exploration, let us develop a formula for this negative exponent case and see what comes of it.

For *substitution rules,* we showed that the similarity dimension is $D_{sim}(X) = \log n / \log s$ for any scheme that replaces a shape with n shapes that are s times smaller. Consider instead the *duplication rule* that repeatedly duplicates a set n times, causing its length to scale by t. The same log–log gradient can be obtained for this new scheme, but the shape it represents is very different. Rather than a finite shape of infinitely fine detail, it is an infinitely large shape with no fine detail. This gradient is the structure's *negative similarity dimension*:

$$D^-_{sim}(X) = \log(n)/\log(t)$$

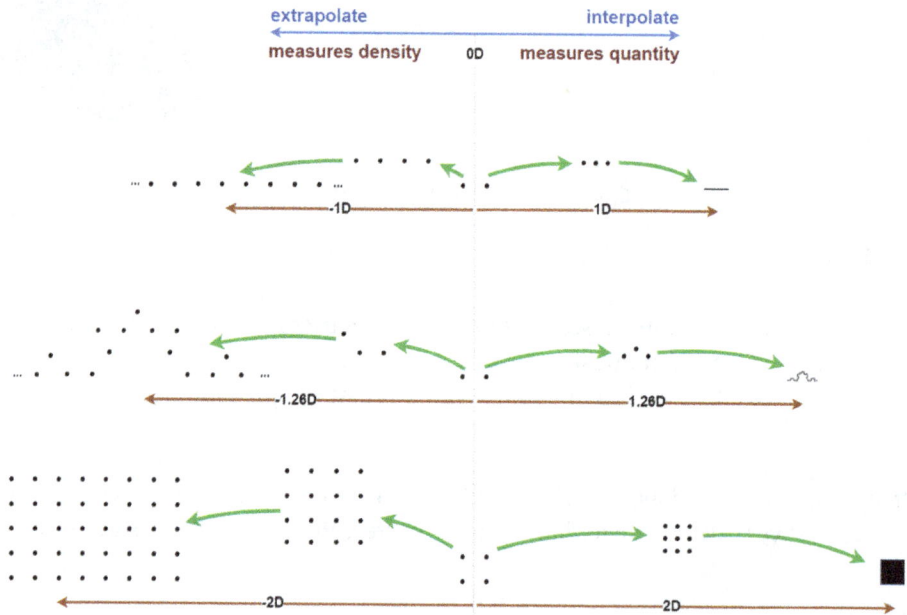

Figure 7.   Examples of iterations 0, 1 and ∞ of the *substitution rule* (rightwards) and *duplication rule* (leftwards) using points as the base shape. The *negative similarity dimension* is shown as a negative value.

These two types of dimension are like duals of each other, the positive similarity dimension describes the scaling law at the finest scales, whereas this *negative similarity dimension* describes the scaling law at the largest scales. Under this pair of replacement schemes, the dual of a line segment is an evenly spaced set of base shapes along an infinite line. In fact, the shapes aren't required to be evenly spaced, but they must tend to a constant density as we look at larger sections. So, this negative 1D structure matches the use of m⁻¹ in everyday usage,[1] it is a structure that has a density of something along a line, such as leaves along a vine.

---

[1] For some different notions of negative dimension, see also [41–43] and Chapter 7.

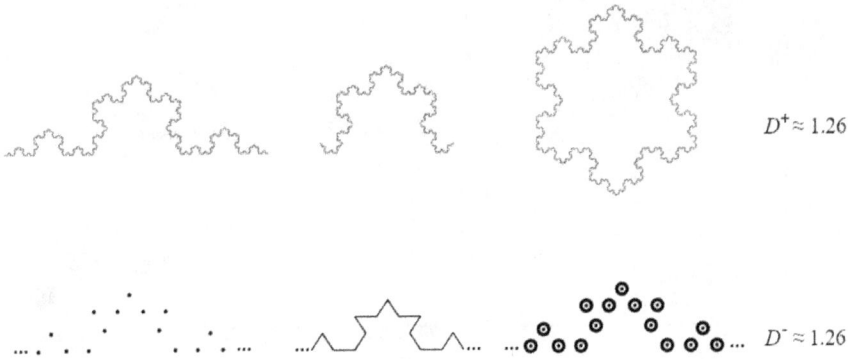

$D^+ \approx 1.26$

$D^- \approx 1.26$

Figure 8.   Positive and negative dimensions. Top: different sections of the *Koch snowflake* have the same positive similarity dimension. Bottom: different base shapes have the same *negative similarity dimension*.

When these infinite structures are composed of points, they have no positive similarity dimension and we can say that they are purely negative dimensional structures by our definition. Equally, finite sized shapes have no *negative similarity dimension*, so we can say that they are purely positive dimensional structures.

But these two dimensions are not mutually exclusive, for example, an infinite 2D lattice of disks has positive dimension 2 and its negative dimension is also 2. This corresponds it being measured by its disk area per square metre, or in units of m^2/m^2.

We can now create a *signed similarity dimension* $D_{sim}^{\pm} = D_{sim} - D_{sim}^-$ to generalise the two types, so the above disk lattice is zero-dimensional. However, due to the non-exclusivity, it is often helpful to write the dimension as a difference, so the above lattice of disks could be described as 2−2D.

A consequence of this *signed similarity dimension* definition is that infinite lines and infinite planes have *signed similarity dimension* 0 since their positive and negative dimensions are both equal. This corresponds to the fact that such shapes can only be measured by counting them, which is a zero-dimensional measure.

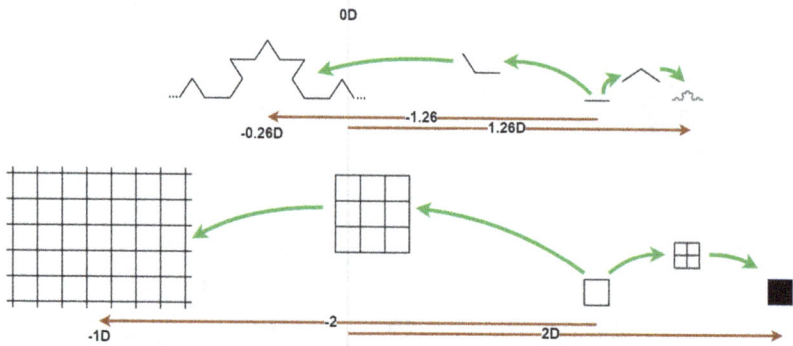

Figure 9.   Examples of iterations 0, 1 and ∞ of the *substitution rule* (rightwards) and *duplication rule* (leftwards) for non-trivial base shapes (a line and a square). The left structures therefore have a positive *and* negative dimensional component, giving a *signed similarity dimension* (left value).

The signed dimension definition above is specific to the *duplication rule* described. For structures built from more general rules, we need a *signed Minkowski dimension*. The positive and negative Minkowski dimensions both calculate the limit in both scale directions, but they differ in the order in which the limits are taken. The positive dimension's inner limit is with respect to reducing resolving width, whereas the negative dimension's inner limit is with a growing bounding window.

Signed Minkowski Dimension

Regardless of the choice of $N(m)$ from Panel 1, we can define a positive and a negative dimension that is valid for unbounded sets by taking the limit with respect to an outer bound window of width w. The *positive Minkowski dimension* is:

$$D_M^+(X) = \lim_{w \to \infty} \lim_{m \to \infty} \frac{\log N(m)}{\log m}$$

and the *negative Minkowski dimension* is:

$$D_M^-(X) = \lim_{m \to \infty} \lim_{w \to \infty} \frac{\log N(m)}{\log w}$$

Giving the *signed Minkowski dimension*:

$$D_M^\pm(X) = D_M^+(X) - D_M^-(X)$$

Like the standard Minkowski dimension, this signed dimension is linear with respect to the *Cartesian product* $A \times B$:

$$D^{\pm}(A \times B) = D^{\pm}(A) + D^{\pm}(B)$$

and with respect to the *Minkowski sum A + B* when it is invertible:

$$D^{\pm}(A+B) = D^{\pm}(A) + D^{\pm}(B)$$

This generalisation of fractal dimension to signed numbers broadens the set of structures that we can quantify. The negative dimensional structures represent lattices, tilings and tessellations, and when the negative dimension is non-integer, they represent more complex, aperiodic structures, including those produced by cellular automata. In each case, the positive dimensional component is used in measuring the *quantity* of structure, and the negative dimensional component is used in measuring the *density* of the structure.

The space-filling curves also find natural representation using signed dimensions, in particular, as 1–2D sets. This avoids the troublesome definitions discussed in Chapter 3. By using a *duplication rule* rather than *substitution rule*, the curve's structure is maintained.

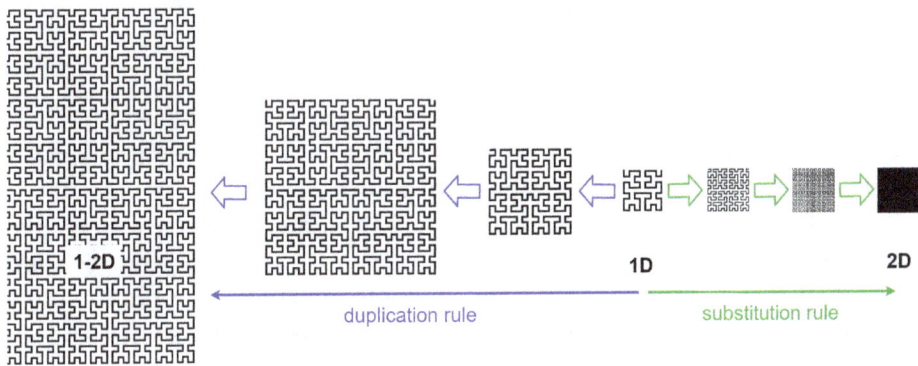

Figure 10.   Hilbert curve iterations. Right: from 1D to a limit as a filled square. Left: from 1D to a 1–2D unbounded set.

The resulting unbounded shape is not space-filling in terms of being a dense set, but it does fill the entire plane with structure, so fills space in the macroscopic sense.

Despite the broader set of structures, signed dimensional shapes still fit within the classification system of Chapter 8. They are just classifying within different scale ranges. Above a particular minimum scale, we can classify the negative dimensional structures, and below a maximum scale, we can classify the positive dimensional structures. These represent two types of *local* scale symmetry. For structures with a positive *and* a negative dimensional component, we must classify each component separately. For instance, we could say that a structure is microscopically a void-sponge, but macroscopically a tree.

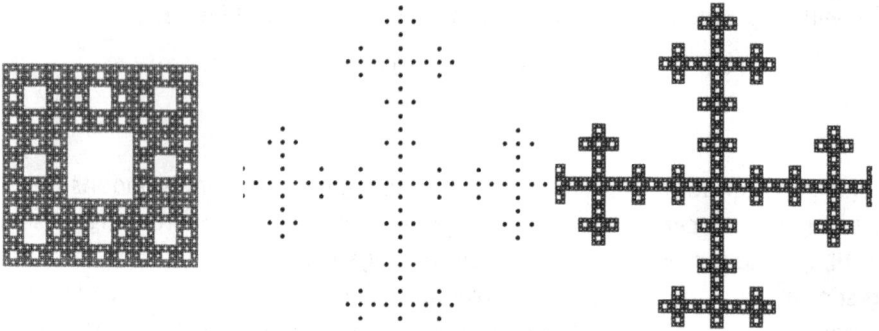

Figure 11.   Classifying signed dimension shapes. Left: void-sponge below a scale limit (1.89D). Middle: void-tree above a scale limit (−1.46D). Right: void-sponge at small scales and void-tree at large scales (1.89–1.46D).

This is really no different to how we would describe other compound structures, such as a structure that is a tree at the top and a sponge at the bottom. The distinction is just by scale rather than location. So, wherever a part of the structure is scale-symmetric, that part can be classified.

★

Looking back at these last two chapters, we can now classify and quantify the diversity of scale-symmetric shapes presented in this book. This includes fractals, non-fractals, static and dynamic structures, and the scale symmetry may be *global* or *local, exact,* or *approximate*. This makes for a concise way to organise this broad landscape of structures. Let us now put the methods into practice and finish the chapter by using these attributes to summarise some well-known shapes and some of the structures presented in this book:

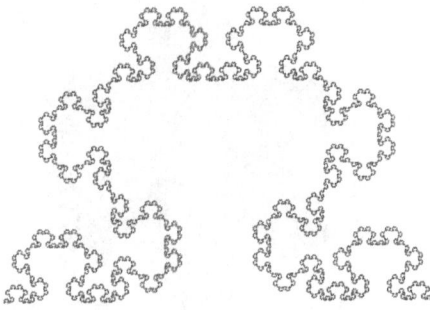

Type: (void-) tree (order 1)
Width: $1\ m$
Length: $1.46\ m^{1.5}$

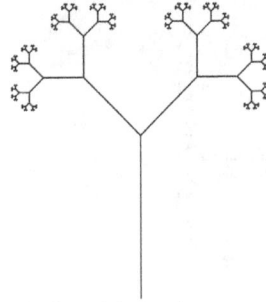

Type: (void-) tree
Trunk height: $1\ m$
Length: $1.44\ m\ log\ m$
Number of leaves: $2.33\ m$

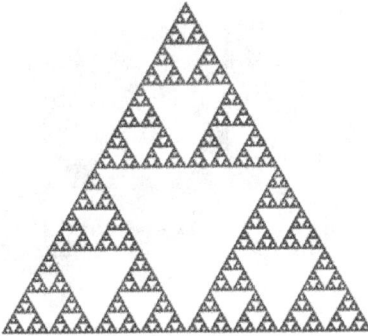

Type: (void-) sponge
Width: $1\ m$
Length: $1.31\ m^{1.59}$

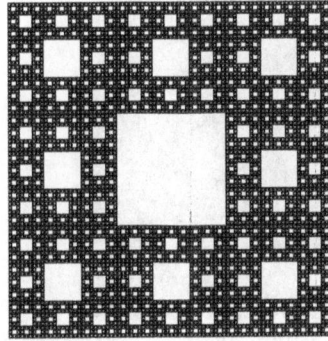

Type: (void-) sponge
Width: $1\ m$
Length: $1.24\ m^{1.89}$

Figure 12.   Examples of measuring non-Euclidean shapes. Coefficients and exponents refer to the normalised Minkowski content and Minkowski dimension, respectively. The "Number of" field refers to the number of points at the object centres. All values are approximate, based on box-counting.

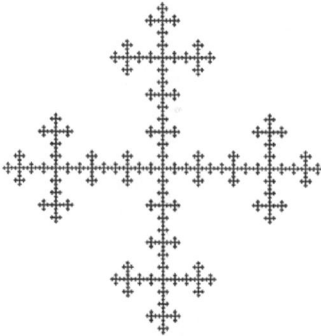

Type: (void-) tree
Width: $1\ m$
Length: $1.56\ m^{1.46}$

Type: (tree-) tree
Width: $1\ m$
Area: $0.69\ m^2$
Perimeter: $1.40\ m^{1.26}$

Type: (tree-) solid
Width: $1\ m$
Area: $1\ m^2$
length of border: $1.96\ m^{1.59}$

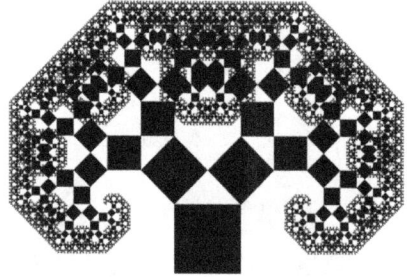

Type: (sponge-) tree
Base square width: $1\ m$
Area: $2.89\ m \log m$

Type: (tree-) tree
Base width: $1\ m$
Area: $0.86\ m^2$
Perimeter: $1.44\ m \log m$
Number of leaves: $2.56\ m$
Branch length: $1.25\ m \log m$
Number of branch points: $3.88\ m$

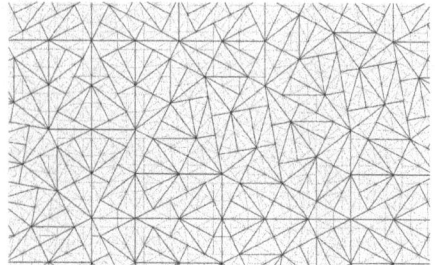

Type: (void-) solid
Shortest edge length: $1\ m$
Number of triangles: $1\ m^{-2}$
Length: $2.62\ m/m^2$
Name: Pinwheel tiling

Figure 12.   (*Continued*)

Cellular Automaton Rule 90

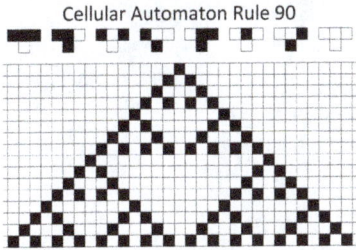

Type: (void-) sponge
Width of pixel: $1\ m$
Number of pixels: $1/3\ m^{-1.585}$
Area: $1/3\ m^{2-1.585}$

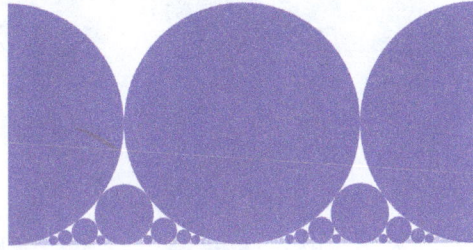

Type: (cluster-) sponge
Height: $1\ m$
Area: $0.87\ m^2/m$
Perimeter: $2.4\ m^{1.169}\ /m$

Type: (sponge-) tree[2]
Height: $1\ m$
Area: $0.60\ m^2$
Perimeter: $1.12\ m^{1.59}$
Number of disks: $0.70\ m^{1.59}$

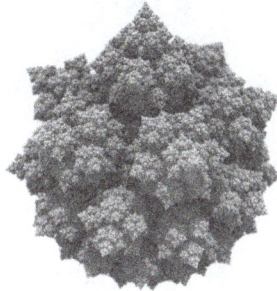

Type: (sponge-) tree[2]
Height: $1\ m$
Volume: $0.21\ m^3$
Surface area: $0.74\ m^{2.56}$
Number of spheres: $0.54\ m^{2.56}$

Figure 12.   (*Continued*)

---

[2] Could we extend the graded classification a step further? If so, then this may be classified more specifically as a cluster-sponge-tree: if it were any thicker, it would be a sponge-tree, any thinner and it would be a cluster.

Chapter 10

Conclusion

The chapters in this book have given an idea of the diversity of mathematical structures that are symmetric under scaling. The examples show the kind of structures that can be built and the ideas behind them. The variety of new structures supports the idea that this field is still open and developing, with many areas left to explore and questions that remain to be answered.

I think there are two reasons why the field is so open. Firstly, as discussed in the introduction, the serious study of geometries with scale symmetry came in the 20th century, which is recent compared to the traditional Euclidean geometries from the time of the ancient Greeks. So, it has simply had less time to mature as a subject. Secondly, scale symmetry is a multidisciplinary subject, which prevents it being a core topic in any one discipline. It is somewhat too playful and undirected for the sciences and maths, too technical for artist courses and requires maths and art skills that a dedicated software engineer might not have. This has delayed its progress until recently, when online communities have catalysed cross-discipline collaborations, and general purpose Graphical Processor Units (GPUs) have accelerated the ability to render and explore in 3D.

Overall, this book has been an exploration for exploration's sake, but there are also some practical uses and connections that can be made to real-world situations, so let's take a look at some of these.

There are about six areas of interest that overlap with the presented topics: science, mathematics, physics, education, programming and art. So, let's briefly discuss each area and then summarise how you can make your own version of the structures in this book before concluding.

Science

As time goes on, it becomes increasingly evident that rules that the mathematician finds interesting are the same as those that nature has chosen.

— Paul Dirac

From the science point of view, the most interesting aspect of scale symmetry is probably its close connection with patterns in nature.

It is well established that many examples in nature show fractal-like self-similarity, such as the clouds, trees and lightning pictured in Chapter 1. There are three plausible reasons for this. Firstly, fractal geometry generalises Euclidean geometry, so it is simply more likely that a structure with symmetries will include some scale symmetry rather than none at all. Secondly, many laws of physical systems operate similarly at multiple scales, as discussed in Chapter 7, the structures that form from these laws reflect the symmetry in the laws themselves. Thirdly, nature often acts over time to optimise some characteristic. For example, rivers roughly minimise the resistance to their flow [34], and the rough shape of coastlines is hypothesised to maximise dissipation of wave energy.

Figure 1.   Optimisation in scale symmetry. Top: rivers roughly minimise their resistance to flow, and rough shorelines may evolve to maximise the dissipation of wave energy. Bottom: large truss structures like the Eiffel tower maximise the strength-to-weight ratio, and bird bones show sponge-like trusses for the same purpose.

Such optimal shapes typically involve non-integer power laws, so are described by fractal geometry.

It is in this third sense that we can understand man-made and evolved scale symmetry, such as the Eiffel tower and bird bone structures, which both roughly maximise strength-to-weight ratio. Another example is fractal antennae, which maximise the length of receiver for their area. So, in this sense, studying scale symmetry may help us to find new optimal designs. Could the 2.5D surface of Chapter 2 optimise heat dissipation? Could the 3D tree of Chapter 4 be an optimal shape to seed new growth on coral reefs? These are just example questions, but they are worth asking because this field of technology is really in its infancy. Most scale-symmetric structures are only practical to produce with the increasing affordability of 3D printing, so there is a lot of future potential for non-Euclidean manufacturing to create more efficient devices.

Maths

From a maths point of view, the importance of scale-symmetric structures is really in their generality and how they connect with other areas of mathematics.

One-dimensional structures with discrete scale symmetry typically look similar under several different scale-and-translate operations. The extreme example of this is the rational numbers a/b for integers a and b, which are symmetric to any scaling or translation by any other rational number. They are a fundamentally scale-symmetric set. So, it is perhaps unsurprising that the rationals can be found hiding in many other scale-symmetric structures. For example, the location of the bulbs in the Mandelbrot set can be mapped to the rational numbers, with the larger bulbs representing the simpler fractions. The Ford circles fractal is even more direct, the circles contact precisely at the rational numbers.

Scale symmetry and the rationals also turn up in the mathematics of music. Two notes sound harmonious when every a oscillations of the first note is b oscillations of the second for integers a and b. This means that the frequency of the second note is a rational multiple (a/b) of the first. By extension, the full set of notes that are mutually harmonious are the set of rational frequencies, where simpler rationals (with smaller a, b) sound most in tune with each other. So, in the Ford circle representation of the rationals shown in Figure 2, chords with frequencies on the larger circles generally sound the most pure. For instance, the largest circles at frequencies 1 and 2 represent an octave chord.

In the common *equal temperament* scale, the notes have *pure scale symmetry* (you get the same set of notes by scaling the frequencies by $\sqrt[12]{2}$), which allows music to be played in any key, consequently the notes don't lie exactly on the resonant frequencies. However, the choice of 12 semitones gives an almost perfect *fourth* and *fifth* on the scale, seen as notes D and E in Figure 2.

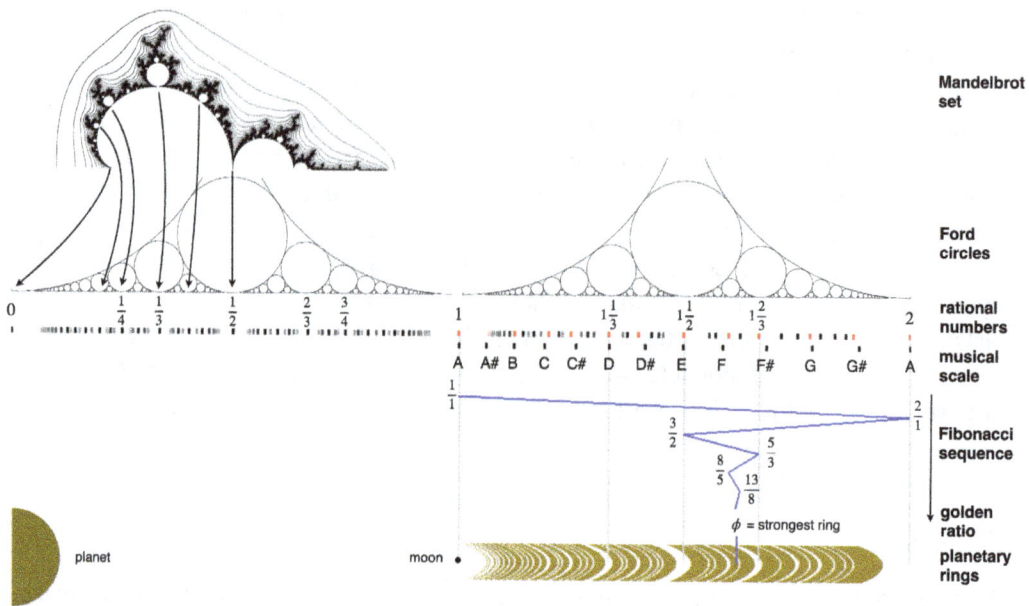

Figure 2.   Example of some connections between scale-symmetric structures and rational numbers: the Mandelbrot set and Ford circle bulbs map to the rationals. The 12-semitone musical scale approximately maps to harmonic frequencies (the nearest simple rational number locations are in red). Planetary rings are unstable (so have gaps) at rational multiples of a moon location.[1] Taking the biggest Ford circle in alternating directions follows a Fibonacci sequence in the rational number's numerator and denominator, this converges to the golden ratio, which equates to the most stable, strongest ring in the planetary system.

The rationals also turn up in dynamic fractal-like systems, such as planetary rings. The gaps in the rings are not random, they are caused by an orbiting moon. For every rational multiple of the moon's orbital period, the rocks at the corresponding orbital distance resonate with the moon's orbit, making them unstable, which results in the gaps. The ring gaps around a planet with a single moon therefore mark out the rational numbers, with the largest gaps being the simplest rationals, which are the strongest resonances.

Physics

On the topic of physics, we demonstrated in Chapter 7 that many natural phenomena can be approximated as systems that are scale-symmetric in space and time, while obeying the laws of physics. These scale-symmetric systems reflect the

---

[1] More correctly, they are rational multiples r of the moon's orbital *period*, so the orbital *radius* of the gaps is $r^{2/3}$ times that of the moon. This is a result of Kepler's third law of planetary motion.

continuous scale symmetry in the underlying laws. In such cases, they are called *scale-invariant* laws.

The key question in identifying scale-symmetric physics is how to account for mass. Since it is a free parameter, we don't know how it should grow with scale. In Chapter 7, we used the assumption that mass grows in proportion to volume on the grounds that density then remains bounded at all scales.

However, a more radical approach is to remove the mass parameter entirely by using a purely geometric model of the physics. We did this in the planetary model of Chapter 7 by swapping the planets with black holes. For non-black-hole objects, there may be other ways to remove the mass parameter and treat the system geometrically using scale-invariant descriptions of the universe,[2] but this ventures into some difficult and speculative areas of theoretical physics.

Nevertheless, the idea that the laws of physics may ultimately be scale-invariant is an enticing one, and it would be a fitting addition to relativistic physics if we finally found that its laws weren't just invariant to translations, rotations, reflections and boosts, but also to scale.

Education

One of the consequences of fractal geometry being a recent field of study is that it remains a topic for higher age groups and hasn't yet filtered down into the education for children.

By age 10, it is common for a child to have learned about shapes like cubes, prisms and pyramids. Why not learn about clusters, trees, sponges, shells and foams as well? The concepts aren't difficult and they represent useful abstractions of natural structures, which is not the case for most Euclidean shapes.

Physical quantities learnt in school are also based on Euclidean roots, they always have an integer dimension. Children learn about lengths, areas and volumes in m^1, m^2 and m^3, so why not also learn to measure things in $m^{1.2}$ or $m^{-2.6}$?

The integer dimensions are ideal for measuring man-made, Euclidean structures, but nature requires fractional dimensions in general. Children should be allowed to learn how to measure the length of coastlines and the surface area of mountains and coral reefs and measure the length of rivers and the veins in our body. These all use fractional dimensions and can be found using the *fractal content* described in Chapter 9, with basic box-counting.

This doesn't just apply to lengths and areas but also to counts. The number of branches on a tree, or number of craters on the moon, or trees in a forest should be counted in $m^{0.2}$, $m^{0.4}$ and so on, rather than in the assumed m^0. These fractional

---

[2] Example topics include *scale-invariant gravity* [36], *conformal gravity* [38] and the *AdS/CFT* correspondence [37].

dimensions are particularly appropriate for counting living things where there are typically many more at the smaller scales, with only a few surviving to full size.

Wouldn't it be great if a child could learn to count starfish in $m^{0.2}$ and measure coral reef surface in $m^{2.3}$ and so could measure the density of invasive starfish on a reef in $m^{0.2}$ per $m^{2.3}$ (which is $m^{-2.1}$). It seems more important than ever that we know how to measure the natural world from a young age.

The structures themselves in this book also have a purpose in education. In teaching a concept, it is useful to have the simplest example at hand, an archetype to demonstrate each idea, and to compare against. Several of the given examples make excellent archetypes, for example, Chapters 2–4 contain archetypal families of surfaces, trees and foams, respectively. There are also some individual shapes that characterise a concept: the 2.5D surface of Chapter 2 is farthest from Euclidean in its dimension, so is a prime example of a fractal surface. The sphere tree of Chapter 4 is the densest such tree without self-overlap, so it is a clear example of a 3D solid with a surface that is everywhere rough. The uptake of 3D printing also allows these examples to be available as physical teaching props.

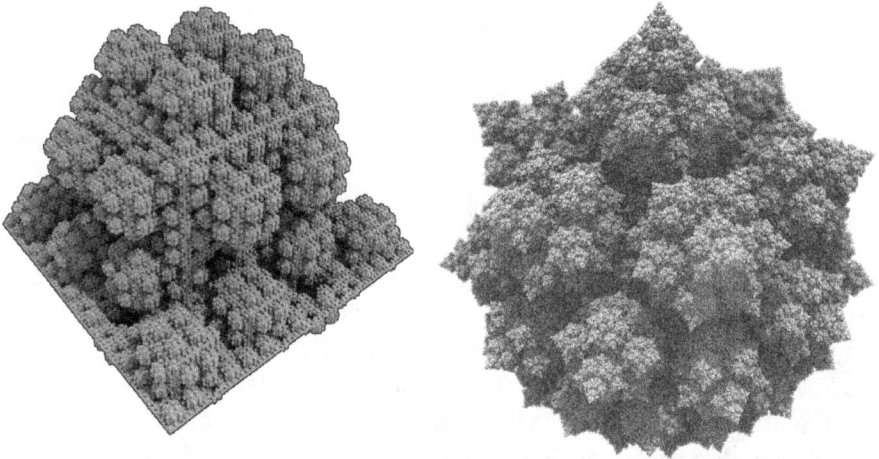

Figure 3.   The archetypal fractal surface (left) and scale-symmetric solid (right) make useful teaching aids.

Programming

From the point of view of the programmer, I think that the covered topics could hold some clues to efficient ways of simulating and rendering. The scale-symmetric automata of Chapter 6 are an interesting template from which to experiment with more general simulations, such as cloth or water simulation. Hierarchical approaches like this can be efficient compared to only simulating the interaction of neighbours at the highest resolution. The multiscale framework is also ideal for immersive environments or virtual worlds, where the detail and update rate should diminish with distance from the camera.

From the rendering side of programming, the "backwards" distance estimate technique described in Chapters 4 and 5 has proven to be very popular for rendering 3D scale-symmetric structures and is the standard method used by modern fractal software.[3] It parallelises easily on Graphical Processing Units (GPUs), it allows deep zooms and requires little memory or data bandwidth. It would be interesting to see how these distance estimate techniques could be generalised to render everyday scenes. A workable system would be a hybrid between the traditional "all data" polygonal rendering and these "all procedural" fractal equations, probably applying the procedural methods at the finer details. This approach has produced some impressively realistic virtual worlds in recent years.[4] Figure 21 in Chapter 6 is an example of this, where a world is built from user-defined voxels, but the 3D structure of the voxels themselves is procedural.

Even for just 2D rendering, fractal concepts can play an important role. Traditionally, each rendered pixel's intensity represents the area of the shape under the pixel. So, how do we render fractals, which have no area? It makes sense for the pixel intensity to be in proportion to the shape's *normalised fractal content* (Chapter 9) instead. This is how the varying bend-angle Koch curve of Chapter 2, Figure 2, was rendered, and it allows anti-aliased image details regardless of the fractal dimension. The same method can be applied to real-world images, which allows image details to remain when downscaled [35], such as power lines or stars in the sky, which otherwise fade away at lower resolutions. This is just another example of an application of fractal techniques to areas outside of geometry.

Art

While there is a lot of depth to the technical study of scale symmetry, ultimately, much of the popularity of this topic is not in the algorithms but in the beautiful structures and patterns that they produce. I know that this is a strong motivating factor for my own interest in the subject.

What exactly makes this field of geometry so visually appealing is a bit of a puzzle. Is it the presence of features at all scales that makes the shapes attractive? Is it the balance of proportions or the visual consistency resulting from the simple underlying formulae? Or is it simply that we have been conditioned to consider fine detail

---

[3] Examples include Fragmentarium and Mandelbulb3D. See back of book for more information on these and related software.

[4] Recent software that combines user-defined large-scale and procedural small details includes http://www.outerra.com/, http://www.procedural-worlds.com/ and https://www.world-machine.com/.

as a sign of effort and expense, giving these structures an association with large gothic cathedrals or elaborate palaces? I don't know, but it is a worthwhile question as it may affect how these pieces of art are developed.

The development of scale-symmetric art, usually called fractal art, is quite a unique process. Unlike traditional art forms such as painting, the user has only a handful of parameters to play with, and these parameters modify the whole structure. Consequently, it is not at all easy to modify or perfect small parts of the structure. This has been an issue in movie special effects companies that have used fractals, as they are not very amenable to local manipulation and so you can't easily remove or move a feature. It is after all the global consistency of the structure that gives it its appeal in the first place. So, the artistic process is closer to that of photography, where choice of subject, lighting, placement and focus are factors that the artist considers.

The following are some examples of variants of the *Mandelbox* shape from Chapter 5, where artists have used lighting effects to bring a different character and ambiance to each piece. One of the more popular variants is called the *Amazing Surface* structure, which is simply a Mandelbox with one axis missing from the box-fold and a small rotation applied at each iteration.

Figure 4.   Scale 2 Mandelbox using structural features as light sources and glow effects to create a futuristic style.

Figure 5. Scale 1.5 Mandelbox with coloured lighting and use of a focal length to help reveal the shape's depth.

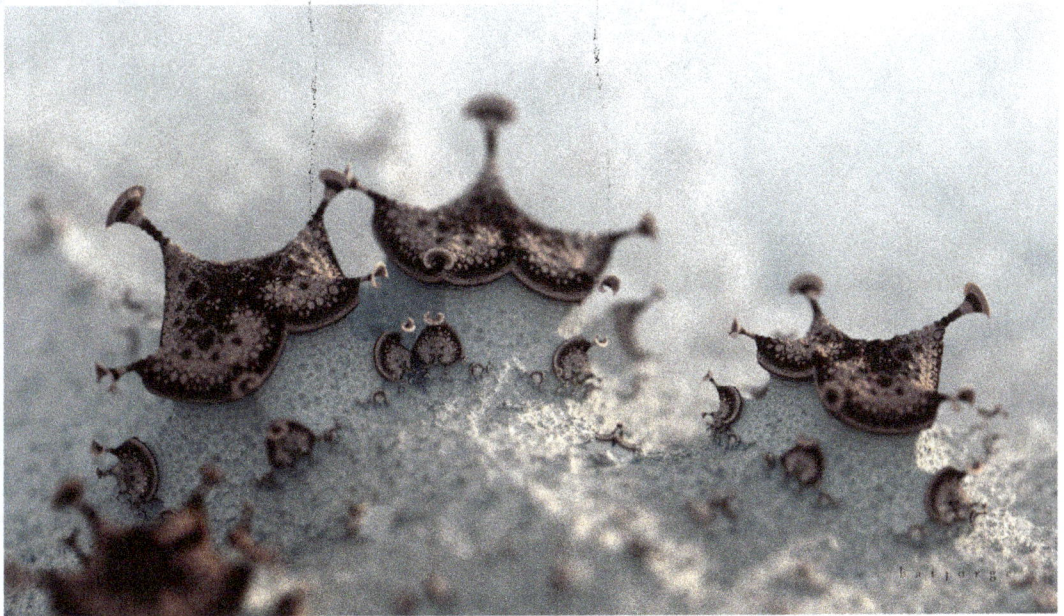

Figure 6. These *Amazing Surface* structures make use of iteration-based colour gradients to reveal surface patterns.

Many of these structures seem to have an innate character to them, and artists often act to emphasise the mood already existing in the piece. It is interesting to consider whether the different sorts of structure in the classification list tend to evoke different feelings. The following four images are cluster-like through to shell-like structures, so they are increasingly self-connected. Does the decreasing freedom to escape out of these structures give them an increasing eeriness? Or does the extra structure give a sense of safety from the outside world? What do you think?

Figure 7.   *Amazing Surface* (Mandelbox variant), with heavy use of focus to suggest a small object.

Figure 8.   Scale 1.5 Mandelbox, with water effect added and use of focus and lighting for dramatic effect.

Figure 9.   These sponge-like *Amazing Surface* structures are reminiscent of a stack of bones or the nest of an unfriendly insect.

Figure 10.   The other-worldly feel of the thin, extruded features in this Mandelbulb are enhanced by the lighting, mist and colour, and the choice of an enclosed cave-like location.

Building the Examples

Regardless of your specialty, or whether your interest is just as a hobby, the best way to understand the examples in this book is to try them out yourself and play around with their formulae. To do this, the source code for the examples in each chapter is linked to on this book's website: https://www.worldscientific.com/worldscibooks/ 10.1142/11219.

Many are dedicated C++ algorithms that can be compiled and run, others are GPU code that is run on third-party rendering software. A full list of the software used is available at the back of the book, but one of the most useful for 3D structures is Fragmentarium. It is free to use and allows you to code your own distance estimation function as well as adjust parameters and lighting through the user interface. For users less interested in modifying code, Mandelbulb3D and Mandelbulber are popular choices for 3D rendering, particularly for the sort of structures in Chapters 4 and 5. All three have been developed with the support of the community at *fractalforums*. I do recommend connecting with communities like this, the breadth of skills available is helpful and they provide a great place to showcase your latest piece of work.

Summary

So, perhaps these notes have given some extra context to the structures and ideas in this book. These ideas reflect the directions I pursued in exploring scale symmetry as a hobby over many years, but there are many other directions that could be taken, the field is broad with much still to discover. And that is the main message of this book — that scale-symmetric geometry is an open and exciting field of study, where new discoveries, new shapes, and new ideas can still be made readily.

With the descriptions in this book and the available source code, I hope that the reader is encouraged to build on these ideas and use them to discover something new. It is a wide world to explore.

Software

The examples in this book can be generated using the dedicated C++ programs available at either of the following sites:

www.worldscientific.com/worldscibooks/10.1142/11219
github.com/TGlad/ExploringScaleSymmetry

This source code is ordered by chapter and can be built and run on the major platforms using a C++ compiler. Most of these programs output images to *.bmp* format, for viewing in a paint program, or they output to *.svg* vector format and can be viewed by dragging them into your web browser. The following external software are also used:

Blender (blender.org) and ArtOfIllusion (artofillusion.org)
Used for the 3D surfaces in Chapter 2. Dedicated C++ programs from the source code link (above) generate the surfaces as triangular meshes. These can then be opened in the above programs for rendering.

Fragmentarium (syntopia.github.io/Fragmentarium)
This is used for 3D structures that have a custom distance estimation function. The 3D grid fractals (Figure 22) in Chapter 2, the 3D images in Chapter 4, and the 3D example classes in Chapter 8 all use this rendering software. The scripts for these functions are available in the source code link.

FractalLab (hirnsohle.de/test/fractalLab)
This GPU web tool is useful for escape-time fractals. It is used to render the tetrahedral Mandelbrot set in Chapter 5 (Figure 6). The required shader for this is available in the source code link.

Ultrafractal (ultrafractal.com)
Mandelbulb3D (andreas-maschke.com/?cat=17)
Mandelbulber (sourceforge.net/projects/mandelbulber)
Used to render the 3D images in Chapter 5, these are specialist fractal rendering tools with a list of built in fractal formulae. The *Mandelbox* formula is called "Mandelbox" in Ultrafractal and Mandelbulber, and "AmazingBox" in Mandelbulb3D. The image in Figure 7 shows the TGladTetra formula and uses Mandelbulb3D. The 3D *Mobius multisets* do not use this software, but are generated from a dedicated C++ program in the source code link.

Image Attributions

Chapter 1

Figure 1. (a) *Sierpinski Triangle* by Beojan Stanislaus used under CC BY-SA 3.0:*https://en.wikipedia.org/wiki/Sierpi%C5%84ski_triangle#/media/File:Sierpinski_triangle.svg.
(c) *Mandel zoom 07 satellite* by Wolfgang Beyer used under CC BY-SA 3.0: https://commons.wikimedia.org/wiki/File:Mandel_zoom_07_satellite.jpg.
(d) *Menger by Baserinia* under the GNU Free Documentation License, Version 1.2, with no Invariant Sections, no Front-Cover Texts, and no Back-Cover Text:*https://commons.wikimedia.org/wiki/File:Menger.png.

Figure 2. (c) Black sun icon by MGalloway, used under CC BY-SA 4.0, modified to include symmetry arrow: https://commons.wikimedia.org/wiki/File:OOjs_UI_icon_sun-ltr.svg.

Figure 6. (b) *Aggregation limitee par diffusion* by Alexis Monnerot-Dumaine under the GNU Free Documentation License, Version 1.2, with no Invariant Sections, no Front-Cover Texts, and no Back-Cover Texts:*https://commons.wikimedia.org/wiki/File:Aggregation_limitee_par_diffusion.png.
(c) *Frozen Spiders Web* by webheathcloseup used under CC BY 2.0 license: https://www.flickr.com/photos/76631347@N00/5209779135.

Figure 7. (a) *White Branches Tree Fractal Silhouette Black* by Max Pixel used under Creative Commons Zero - CC0: https://www.maxpixel.net/White-Branches-Tree-Fractal-Silhouette-Black-1438260.
(b) *snow-landscape-mountains-nature-59106* by liszt yu, free to use: https://www.pexels.com/photo/snow-landscape-mountains-nature-59106/.
(c) *Cumulus cloud* by Jon Candy used under CC BY-SA 2.0: https://www.flickr.com/photos/joncandy/9421862301.
(d) *water_sound_waves_seagull* under Creative Commons CC0: https://pxhere.com/en/photo/764987.

Figure 9. Based on data collated in: en.wikipedia.org/wiki/Asteroid#Characteristics.

Figure 10. *Britain-fractal-coastline-combined*, originals made by Avsa mixed by Acadac, used under CC BY-SA 3.0: https://commons.wikimedia.org/wiki/File:Britain-fractal-coastline-200km.png#/media/File:Britain-fractal-coastline-combined.jpg.

Chapter 2

Figure 13. Textured using image by Tim Holt, source: https://lemma.forestry.oregonstate.edu/images/gnnviz/fakesat_sample2.jpg.

Chapter 3

Figure 3. *Pythagoras tree 1 0 8 13 prism* by Guillaume Jacquenot used under CC BY-SA 1.0, 2.0, 2.5 and 3.0: https://he.wikipedia.org/wiki/קובץ:Pythagoras_tree_1_0_8_13_prism.svg.

Figure 11. (a) *Peanocurve* by António Miguel de Campos used under GNU Free Documentation License, with no Invariant Sections, no Front-Cover Texts, and no Back-Cover Texts:*https://commons.wikimedia.org/wiki/File:Peanocurve.svg.
(b) *S2 Hilbert Curve L4 All* by Vahram Mekhitarian used under Creative Commons Attribution-Share Alike 3.0 Unported:* https://commons.wikimedia.org/wiki/File:S2_Hilbert_Curve_L4_All.jpg.

Figure 14. *Peanocurve* by António Miguel de Campos used under GNU Free Documentation License, with no Invariant Sections, no Front-Cover Texts, and no Back-Cover Texts. Modified to add solid square at right: https://commons.wikimedia.org/wiki/File:Peanocurve.svg.

Chapter 4

Figure 2. (a) *Asteroid and it's Pile of Rubble* by Kevin Gill used under CC BY 2.0: https://www.flickr.com/photos/kevinmgill/15148296739.
(b) *Water drops* by Dima Bushkov used under CC BY 2.0: https://www.flickr.com/photos/bushkov/4197513195.

Figure 3. (a) and (b) Credit to JAXA, University of Tokyo, Kochi University, Rikkyo University, Nagoya University, Chiba Institute of Technology, Meiji University, University of Aizu, AIST.

Figure 4. (a) *Pebbles on Findhorn Beach* by Andrew Urquhart used under CC BY-SA 2.0: https://www.flickr.com/photos/68958307@N00/4891120213.
(b) *Pebbles on the beach at Gunwalloe* by Tim Green used under CC BY 2.0: https://www.flickr.com/photos/atoach/2408459510.

Chapter 5

Title: *Mandel_zoom_06_double_hook* by Wolfgang Beyer used under GNU Free Documentation License, Version 1.2, with no Invariant Sections, no Front-Cover Texts, and no Back-Cover Texts: https://commons.wikimedia.org/wiki/File:Mandel_zoom_06_double_hook.jpg.

Figure 1. *Mandel zoom 00 mandelbrot set* by Wolfgang Beyer used under CC BY-SA 3.0 Unported, 2.5, 2.0 and 1.0 generic licenses. Modified with gradient change and axes added:* https://en.wikipedia.org/wiki/File:Mandel_zoom_00_mandelbrot_set.jpg.

Figure 3. (b) *Parabolic julia set c = −0.75* by Adam majewski used under CC BY-SA 3.0: https://commons.wikimedia.org/wiki/File:Parabolic_julia_set_c%3D-0.75.png.
(c) *Parabolic Julia set for internal angle 1 over 30* by Adam majewski used under CC BY-SA 4.0: https://commons.wikimedia.org/wiki/File:Parabolic_Julia_set_for_internal_angle_1_over_30.png.
(d) *Julia set of the quadratic polynomial f(z) = z^2 − 1.12 + 0.222i* by Lasse Rempe-Gillen used under CC BY 3.0: https://commons.wikimedia.org/wiki/File:Julia_set_of_the_quadratic_polynomial_f(z)_%3D_z%5E2_-_1.12_%2B_0.222i.png.

Figure 4. Images courtesy of Krzysztof Marczak: www.mandelbulber.com.

Figure 7. (a) Image by Luca: https://fractalforums.org/mlist/darkbeam_34.
(b), (d) and (e) Courtesy of pupukuusikko: https://www.deviantart.com/pupukuusikko.
(c) Courtesy of Johan Andersson: https://www.deviantart.com/mandelwerk.

Figure 11. (b) By Krzysztof Marczak: www.mandelbulber.com.

Figure 21. (b) By Sam Derbyshire:*math.ucr.edu/home/baez/roots/beauty.pdf.

Chapter 6

Gliders: Images courtesy of Alan Dorin, from the article: *A Framework For Understanding Generative Art.*

Chapter 8

Title: (a) *Froth and Bubble* by Michael Coghlan used under CC BY-SA 2.0: https://www.flickr.com/photos/mikecogh/7567416638.
(c) *New Sponge* by NOAA Ocean Exploration used under CC BY-SA 2.0: https://www.flickr.com/photos/oceanexplorergov/33433221746.
(e) *Phytokarst on aragonitic calcarenitic eolianite limestone* by James St. John used under CC BY 2.0: https://www.flickr.com/photos/jsjgeology/16135327980.

Figure 4. **Tree-sponge:** Single frame of *Quadriccross* by Akarpe used under CC BY-SA 3.0: https://commons.wikimedia.org/wiki/File:Quadriccross.gif.
Sponge-sponge: *Harriss Spiral* by Edmund Harriss used under CC BY-SA 3.0: https://www.theguardian.com/science/alexs-adventures-in-numberland/2015/jan/13/golden-ratio-beautiful-new-curve-harriss-spiral#img-1.
Void-solid: see Chapter 3, Figure 11, and Chapter 9 Figure 11.

Figure 5. **Tree-void:** *Taula Fractal- Fractal table at DHUB* by Kippelboy used under CC BY-SA 3.0: https://commons.wikimedia.org/wiki/File:Taula_Fractal-_Fractal_table_at_DHUB.jpg.
Fractal-plant by Sakurambo used under CC BY-SA 3.0: https://de.wikipedia.org/wiki/Datei:Fractal-plant.svg.
Void-tree: created by Jacqueline Yen, Kevin Chu, and Chris H. Rycrof: http://people.seas.harvard.edu/~chr/research/3d-dla/.
Shell-tree: image by Johannes Middek, from the Research Institute for Symbolic Computation at JKU: https://www3.risc.jku.at/education/courses/ws2016/cas/landscape.html.
Void-sponge: *Koch Curve in Three Dimensions ("Delta" fractal)* by Eric Baird used under CC BY-SA 4.0: https://commons.wikimedia.org/wiki/File:Koch_Curve_in_Three_Dimensions_(%22Delta%22_fractal).jpg.
Single frame of *Menger sponge diagonal section* by Cmglee used under CC BY-SA 4.0: https://commons.wikimedia.org/wiki/File:Menger_sponge_diagonal_section.gif.
Void-shell: Frame of *KochCube Animation Gray* by Guillaume Jacquenot used under CC BY-SA 3.0 Unported, 2.5, 2.0 and 1.0 Generic: https://commons.wikimedia.org/wiki/File:KochCube_Animation_Gray.gif.
(and for tree-solid) *Koch surface 3* by Robert Dickau used under CC BY-SA 3.0: https://commons.wikimedia.org/wiki/File:Koch_surface_3.png.
Tree-shell: *Fractal rhombic dodecahedron* by fdecomite used under CC BY 2.0: https://www.flickr.com/photos/fdecomite/5372394711/in/photostream/.

Void-solid: *Moore3d-step3* by Robert Dickau used under CC BY-SA 3.0: https://commons.wikimedia.org/wiki/File:Moore3d-step3.png.
3D H-fractal by Phyhoubo used under CC BY-SA 3.0: https://commons.wikimedia.org/wiki/File:3D_H-fractal.png.
Cluster-solid: *Apollonian spheres* by Paul Bourke used under CC BY-SA 3.0: https://commons.wikimedia.org/wiki/File:Apollonian_spheres.jpg.

Figure 8. (a) *Expl0967 (9734070707)* from NOAA Photo Library used under CC BY-2.0: https://commons.wikimedia.org/wiki/File:Expl0967_(9734070707).jpg.
(b) *Expl5561 (9734255741)* from NOAA Photo Library used under CC BY-2.0: https://commons.wikimedia.org/wiki/File:Expl5561_(9734255741).jpg.
(c) *Expl5585 (9734255031)* from NOAA Photo Library used under CC BY-2.0: https://commons.wikimedia.org/wiki/File:Expl5585_(9734255031).jpg.
(d) *Expn4440 (27864969991)* from NOAA Photo Library used under CC BY-2.0: https://commons.wikimedia.org/wiki/File:Expn4440_(27864969991).jpg.

Chapter 9

Figure 10. Derivative of *S2 Hilbert Curve L4 All* by Vahram Mekhitarian used under Creative Commons Attribution-Share Alike 3.0 Unported: https://commons.wikimedia.org/wiki/File:S2_Hilbert_Curve_L4_All.jpg.

Figure 11. Pinwheel: *Pinwheel 3* by Levochik used under GNU Free Documentation License, Version 1.2, no Invariant Sections, no Front-Cover Texts, and no Back-Cover Texts. And under CC BY-SA 3.0 Unported, 2.5, 2.0 and 1.0 Generic: https://commons.wikimedia.org/wiki/File:Pinwheel_3.jpg.

Chapter 10

Figure 1. (a) Image by NASA/GSFC/LaRC/JPL, MISR Team.
(c) *Eiffel Tower, Paris 7th 002* by Moonik used under CC BY-SA 3.0: https://commons.wikimedia.org/wiki/File:Eiffel_Tower,_Paris_7th_002.JPG.
(d) Swan leg bone image, courtesy of Karen Harvey: www.karen-harvey.co.uk.

Figure 2. Derivative work including *LCMM* by Adam Majewski used under GNU Free Documentation License Version 1.2, with no invariant sections, no front-cover texts and no back cover texts. Also under CC BY-SA 4.0, 3.0, 2.5, 2.0 and 1.0: https://commons.wikimedia.org/wiki/File:LCMM.jpg.

Figures 4, 5, 8 and 10. Images courtesy of Krzysztof Marczak: www.mandelbulber.com, www.deviantart.com/krzysztofmarczak.

Figures 6, 7 and 9. Images courtesy of batjorge:* www.fractal.batjorge.com/.
* Thumbnails also appear in the Glossary of Shapes.

Glossary

Glossary of Shapes

Amazing surface A variant of the *Mandelbox* using only a two-axis box-fold and one rotation per iteration.

Apollonian gasket A void-sponge circle packing, Frederick Soddy 1937.

Buddhabrot set A multiset tracing the iterations of the Mandelbrot set, Melinda Green 1993.

Cantor dust/cantor's (middle-third) set A void-cluster set of points, Georg Cantor 1883.

Crumpled surface fractal A fractal surface of any dimension between 2 and 3, which is a developable surface, so is a crumpled but unstretched version of a flat surface. Overlap occurs at all fractal dimensions more than two, but is minor until approximately 2.3D.

Diffusion limited aggregation A stochastic void-tree set, T.A. Witten and L.M. Sander 1981.

Dragon curve   A space-filling curve (void-solid), described by Martin Gardner 1967.

Ford circles   A cluster-sponge of contacting disks, Lester Ford 1938.

Ford spheres   3D cluster-sponge of contacting balls, extension of Ford circles, Ford 1938.

Fudgeflake   A space-filling curve (void-solid), W.H. Freeman & Co. 1983.

Greek cross fractal   A space-filling tree (void-solid).

Hexaflake   A void-sponge, originating approximately 1990.

Hilbert curve   A space-filling curve (void-solid), David Hilbert 1891.

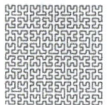

Kleinian fractal   A fractal using transforms within a Kleinian group. Popularised by Mumford, Series, Wright 2002.

Koch curve   A simple fractal curve, Helge Von Koch 1904.

Koch snowflake   A closed fractal curve, Helge Von Koch 1904.

Levy C curve   An overlapping simple fractal curve, Ernesto Cesàro 1906.

Littlewood fractal   A multiset that includes generalisations of the dragon curve within, Dan Christensen 2006.

Mandelbox   A type of 3D Mandelbrot set defined by a combination of box-folds, ball-folds and a dilation. The Julia sets of this shape are *shape-preserving*.

(the) Mandelbrot set	A 2D cluster-tree with border of Hausdorff dimension 2, Benoit Mandelbrot 1980.	
Menger/Sierpinski carpet	A void-sponge built by repeatedly removing middle-third square, Waclaw Sierpinski 1916.	
Menger sponge	A 3D void-sponge built by repeatedly removing middle-third cube, Karl Menger 1926.	
Mobius multiset	A multiset generated as the number of branches that remain within a threshold when applying a pair of *Mobius transformations* at each point in the tree.	
Peano curve	A space-filling curve (void-solid), Giuseppe Peano 1890.	
Pentadentrite	A void-tree with five-fold symmetry, Gerald Edgar 1990.	
Pythagoras tree	A recursive tree structure of square blocks, Albert Bosman 1942.	
Pyramidal surface	A fractal surface of any dimension between 2 and 3, which is generated by a substitution rule that replaces each triangle with six smaller triangles, of which four form a pyramid. Overlap occurs for fractal dimension greater than approximately 2.3.	
Semi-dimensional shapes	Fractals with fractal dimension exactly half way between their topological dimension and the dimension of the space they are embedded in.	
Sierpinski triangle	A void-sponge built by repeatedly removing an equilateral triangle, medieval origins.	
Tridendrite	A void-tree with three-fold symmetry.	

Twin tiles	The name of a pairs of triangles A and B, which can be arranged to form a larger version of A and to form a larger version of B, under a reflection.
Vicsek fractal	A void-tree with square symmetry, Tamás Vicsek 1992.
Weierstrass function	A fractal function related to the Koch curve, Karl Weierstrass 1872.

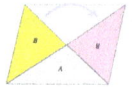

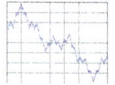

Glossary of Terms

Apollonian problem	To find the circle tangent to three other circles in a plane.
Approximate scale symmetry	*Scale symmetry* where the shape is visually or numerically similar to itself at multiple scales or under shape-preserving transformations.
Behaviour	As used: an $n+1$ dimensional scale-symmetric structure. A behaviour is therefore a scale-symmetric structure that moves and changes with time, where smaller parts change over a proportionally smaller time period.
Boost	The coordinate change representing a change in velocity. In Special Relativity this is a Lorentz boost, or Lorentz transformation.
Box-counting dimension	Type of *fractal dimension*, defined as $\lim\limits_{m\to\infty} \frac{\log N(m)}{\log m}$ where $N(m)$ is the number of pixels of width m^{-1} that cover the set.
Cartesian product	The operation $A \times B$ giving the set of pairs: $\{(a, b)\,|\,a \in A$ and $b \in B\}$.
Cellular automaton	Any rule-based system on a tiling of cells whereby the change in state of each cell is a fixed function of the state of it and its neighbouring cells.
Cluster	A recursive, disconnected and scale-symmetric set of regions or blobs.
Complex fractal dimension	The complex valued generalisation of *fractal dimensions* where the imaginary part characterises the oscillations in the growth curve.

Conformal transformation	A transformation that preserves angles such that small spheres remain spheres and small shapes are not reflected.
Conway's Game of Life	A cellular automaton on a square grid of live/dead cells, whereby only live cells with two or three live neighbours survive, and dead cells with three live neighbours become live.
Co-similarity	A set of distinct shapes such that there is an arrangement of the set that produces each of the shapes, at a larger size. Consequently, each shape in the set will contain the other shapes within it.
CPT symmetry	A symmetry of nature on the product of charge, parity (left or right handedness) and time direction. If you negate any two the system will act the identically.
Diamond-square algorithm	A method that generates a rough, mountainous surface by taking every square in a lattice to generate a mid-point which is offset by a random value in proportion to the square width. In plan view, this generates a new square lattice rotated by 45°, so the process is then repeated.
Distance estimation function	A function that takes any 3D point in space and returns an estimate of the distance from that point to the closest point on the object or scene being rendered.
Double-cover	A transformation on a set is a double-cover when there are two input points for every transformed point. For example, the transformation z^2 is a double-cover on the complex plane.
Duplication rule	A structure generation method whereby the shape is recursively duplicated and the copies transformed in a configuration relative to one another. This rule generates structures of infinite extent at its limit.
Dynamic	A system that considers the velocities, accelerations, forces, torques and masses of bodies.
Embedded in	As used: a shape that is in Euclidean space such that no two points in the shape occupy the same point in the Euclidean space. A shape that has been embedded in a Euclidean space is a subset of the Euclidean space.
Equal temperament scale	A common musical scale whereby each consecutive semitone is a constant multiple of the frequency. Used by pianos and instruments that can change key.

Escape time fractal	A generation method whereby the points in the set are distinguished by whether or not they diverge to infinity under repeated iteration of a chosen function.
Exact scale symmetry	*Scale symmetry* that results in a mathematically identical copy of the shape.
Euler number	The number of disconnected parts in a shape minus the number of holes.
Fourth/fifth (music)	In just tuning, a fourth is a 4/3 higher frequency and a fifth is a 3/2 higher frequency than the base note.
Fractal	A subset of Euclidean space with fractal dimension greater than its topological dimension. This means that its coverage of space grows faster with respect to increasing resolution than its smooth counterpart. 2D fractals cannot have area, nor can 3D fractals have volume, so they can be thought of as "thin" *or* "void" scale-symmetric structures. Common usage of "fractal" is often broader, describing any self-similar object or pattern.
Fractal content	The "amount" of a shape. This value generalises the notion of count, length, area and volume to non-integer dimensional shapes. The "normalised fractal content" is scaled such that the Euclidean primitives — lines, squares and cubes — have their usual values of length, area and volume.
Fractal curvature	The fractal content of one of the *intrinsic curvatures* of a set.
Fractal dimension	General term for a number of functions on a shape that extract the exponent of the rate of its coverage of space with respect to resolution. In all cases, a point, line, square and cube have fractal dimensions 0, 1, 2 and 3, respectively, but they can differ slightly for intermediate dimensions.
Fractal flames	A type of Iterated Function System where non-shape-preserving transformations are included and the path of each iteration is rendered.
Fractal scaling	The fractal dimension of one of the *intrinsic curvatures* of a set.
Fractal forums	Forum website for fractal art, mathematics and programs: www.fractalforums.org.

Frobenius theorem	States that there are only three associative division algebras of finite dimension: the real numbers, complex numbers and quaternion numbers.
Galaxy filament	A long and narrow string of thousands of galaxies, as observed at large scales.
Galaxy wall	A large and relatively flat plane of thousands of galaxies, as observed at large scales.
Glider	A group of cells in a cellular automaton which evolve through a repeating pattern that shifts across the tiling.
Graph of a function	The set of values $(x, f(x))$ over the entire domain of a function f. The graph of a $\sin(x)$ function is a sine wave curve in 2D.
Hausdorff dimension	A form of fractal dimension introduced by Felix Hausdorff. This is frequently equal to the box-counting dimension, but is sometimes less, for instance, the set of rational numbers in $(0,1)$ has Hausdorff dimension 0 rather than 1.
Immersed in	As used: a shape that is locally embedded in Euclidean space, but where more distant areas of the shape may overlap in the Euclidean space. A shape immersed in Euclidean space is not a subset of this space, but its points could be represented by a multiset.
Image of a function	The set of values of the function over its entire domain. The image of a $\sin(x)$ function is the interval $[-1,1]$.
Intrinsic curvatures of a set	Principle characteristics of a set, in 2D they are as follows: the area, the perimeter and the number of separate regions minus the number of holes.
Iterated function system (IFS)	A general fractal generation method that requires iterating a function, typically this refers to substitution rule, multiplication rule and also chaos game generation methods.
Kleinian group	A discrete subgroup of the set of Mobius transformations.
Kinematic	A system that considers the velocities and accelerations of bodies but not masses.
Limit set	The set taken at the limit of a repeated set of multiple transformations. Typically, one starts with a single point and applies n transformations to repeatedly multiply the set size by n.

Liouville's theorem	This theorem of conformal mappings states that any conformal mapping in more than 2D space must be composed of translations, rotations, dilations and orientation-preserving sphere inversions. These are the *Mobius transformations.*
Local scale symmetry	*Scale symmetry* within a scale range, such that the symmetry is lost either below a certain scale, or above a certain scale, or both.
Measure (mathematics)	Assigns a number to any subset of a set, representing its size, such as count, length, area, volume etc. It has a stricter definition than content (as in Minkowski content), in particular it must be countably additive.
Minkowski content	$\lim\limits_{m\to\infty} \frac{N(m)}{m^D}$ where $N(m)$ is the *Minkowski number* and D is the Minkowski dimension of the shape.
Minkowski number	$N_M(m)$ in $\mathbb{R}^n$ is m^n times the n-dimensional volume of all points within radius $r = m^{-1}$ of the shape.
Minkowski spacetime	One time and three space coordinates whereby a change in velocity is a Lorentzian transformation rather than the traditional Galilean transformation.
Minkowski sum	The set operation $A + B$ which equals $\{a + b \mid a \in A$ and $b \in B\}$.
Minimal surfaces	Surfaces that locally minimise their area, such as soap film. These surfaces have zero mean curvature.
Mobius transformation	Transforms circles (and lines) to circles (and lines). These transformations can always be composed of translations, rotations, dilations (uniform scaling) and reflected sphere inversions.
Multiple scale symmetry	Symmetry under more than one shape-preserving transformation.
Multiplication rule	A recursive structure generation method whereby smaller and transformed copies of a base shape are added recursively to a set.
Multiset	Also called a bag, a multiset is a set that accounts for multiple instances of the same element. This can be structured as a set of elements together with their "multiplicity," a natural number defining how many of each element.

Negative Minkowski dimension	Proposed *fractal dimension*, defined as $\lim\limits_{m\to\infty} \lim\limits_{w\to\infty} \frac{\log N(m)}{\log w}$ for a counting function $N(m)$, resolution m and window size w.
Negative similarity dimension	Proposed *fractal dimension*, defined as $\log n/\log t$ for a duplication rule that repeatedly duplicates a set n times, scaling its length by t.
Plateau's laws	Describe the geometry of soap film as follows: (a) made of unbroken smooth surfaces, (b) constant mean curvature for any surface, (c) surfaces meeting at 120° angles and (d) edges meeting as the edges of a radial tetrahedron, angle $\mathrm{acos}(-\tfrac{1}{3})$.
Positive Minkowski dimension	Type of *fractal dimension*, defined as $\lim\limits_{w\to\infty} \lim\limits_{m\to\infty} \frac{\log N(m)}{\log m}$, where $N(m)$ is a counting function, typically the Minkowski number. m is the resolution and w is the window size.
Power law	A law describing how a parameter p varies with scale coordinate m, of the form $p = cm^d$ for some coefficient c and exponent d. For example, the volume of a cube scales with m^3, where m is the side length, so it follows a power law.
Pure scale symmetry	Symmetry under only a uniform scaling (dilation) and no other transform.
Radial tetrahedron	A 3D shape composed of six triangular planes, each with two corners at two different vertices of a tetrahedron and the other corner at the tetrahedron centre.
Rational space-filling structure	As used: a subset of Euclidean space that has no finite gaps or dense regions, such as the set of vectors that have rational valued components.
Relativistic	A system that has no observable absolute positions, orientations or velocities.
Riemann sphere	A representation of the complex plane (plus a point at infinity) as a unit sphere centred at the origin. Sphere coordinates map to the complex coordinates through a stereographic projection.
Scale invariance	The invariance of physical laws or equations under scaling of the length dimensions, and sometimes under scaling of other variables, such as mass or energy. Normally, this refers to invariance under continuous scaling.

Scale-symmetric automata	An automaton that is scale-symmetric in space and time, which is composed of layers of grids at power of two resolutions. Each cell in a layer updates its state as a function of the neighbouring cells in that layer and its adjacent layers.
Scale-symmetric billiards	A recursive distribution of disks or balls moving with equal speed and energy conserving collisions. The problem is whether there is a simple recursive description of their trajectories.
Scale symmetry	The branch of geometry concerned with shapes that are identical under at least one shape-preserving transformation that includes scaling. Its strict interpretation means that transforming the shape by a combination of reflections and conformal transformations gives an identical copy of the original shape.
Schwarzschild black hole	The simplest physical model of a black hole. One that has no angular momentum or charge in an otherwise empty universe that tends to Minkowski spacetime with distance.
Self-similarity	A shape which contains parts that are approximately or exactly similar to the whole shape. A stricter definition is a shape that is composed of a finite number of smaller exact copies of that shape. The meaning is very similar to that of scale symmetry.
Shape-preserving	A transformation under which small spheres remain spheres. This includes conformal transformations together with reflections. A fold transformation is shape-preserving everywhere apart from the fold line itself.
Signed Minkowski dimension	The difference between the positive and negative Minkowski dimensions. Unbounded lines, regions, volumes and unbounded fractals all have signed dimension 0.
Signed similarity dimension	The difference between *similarity dimension* of a shape and its *negative similarity dimension*. Unbounded lines, regions, volumes and unbounded fractals all have signed dimension 0.

Similarity dimension	Type of fractal dimension, defined as log n/log s for any scheme that repeatedly replaces a shape with n shapes that are s times smaller along a length.
Space-filling curve	A function of a single variable where the range (or image) of the function fills the space it is occupying.
Sphere inversion	A transformation that replaces the length of vectors with their reciprocal length. Loosely, this "reflects" the space around the sphere surface. An orientation-preserving sphere inversion also includes one reflection.
Spherical tetrahedron	A sphere composed of four equally shaped sphere segments, where each is the "inflated" face of a tetrahedron.
Statistical scale symmetry	Similarity of a shape under shape-preserving transformations only on average. The principle statistics of the shape (such as standard deviation, and skew) are invariant under the transformation.
Stereographic projection	A projection from a sphere onto a plane. For the *Riemann sphere* onto the complex plane, the complex value at sphere coordinate x, y, z is $(x + iy)/(1 - z)$.
Substitution rule	A fractal generation method by repeatedly applying a rule whereby one shape is substituted by multiple smaller and transformed copies of that shape.
The chaos game	A generation method by which a set of transforms are repeatedly chosen at random and applied to a point in space, the iterations converge to a set of points (a limit set). Usually, these transforms act to move the point a percentage of the way to one of a number of corners of a shape. When the shape is an equilateral triangle, it generates the Sierpinski triangle.
Topological dimension/Lebesgue covering dimension	Formalises the definition of points as 0D, curves as 1D, surfaces as 2D, etc. One can loosely define it as one less than the largest number of overlapping regions that cover the shape, while minimising overlap and region size. For example, a line can be covered by regions that only overlap in pairs, so its topological dimension is 1.
Turing complete	An automaton that can emulate a Turing machine, which is a universal computer. A Turing-complete system can process any data that a computer can.

Upscaling automata   Automata that generate increasingly higher resolution versions of a grid of cells as a function of the local neighbourhood of each cell in the lower resolution layer.

Yardstick method   A form of *fractal dimension* defined as $1 + \lim\limits_{m \to \infty} \frac{\log L(m)}{\log m}$, where $L(m)$ is the length of the curve measured with rulers of length m^{-1}.

References

[1] H. V. Koch, "On a continuous curve without tangents constructible from elementary geometry," in *Classics on Fractals*, Reading, Massachusetts, Addison-Wesley, 1993, pp. 25–45.

[2] M. Barnsley, *Fractals Everywhere*, Boston, MA: Academic Press, 1988.

[3] A. Lindenmayer, "Mathematical models for cellular interaction in development, Parts I and II," *Journal of Theoretical Biology,* vol. 18, pp. 280–315, 1968.

[4] P. Prusinkiewicz, "L-systems: from the theory to visual models of plants," in *Proceedings of the 2nd CSIRO Symposium on Computational Challenges in Life Sciences*, Melbourne, 1996.

[5] B. B. Mandelbrot, *The Fractal Geometry of Nature*, vol. 173. New York: WH Freeman, 1983.

[6] D. R. Davis *et al.*, "Collisional evolution of small-body populations," in *Asteroids III*, UAPress, 2002, p. 545.

[7] G. West *et al.*, "A general quantitative theory of forest structure and dynamics," *Proceedings of the National Academy of Sciences of the United States of America,* vol. 106, no. 17, pp. 7040–7045, 2009.

[8] L. F. Richardson, "The problem of contiguity: An appendix to statistics of deadly quarrels," *General Systems Yearbook,* vol. 6, p. 139, 1961.

[9] B. B. Mandelbrot, "How long is the Coast of Britain? Statistical self-similarity and fractional dimension," *Science,* vol. 156, no. 3775, pp. 636–638, 1967.

[10] M. Hazewinkel, *Encyclopedia of Mathematics*, Springer Science+Business Media B.V./ Kluwer Academic Publishers, 2001. [Online]. Available: encyclopediaofmath.org/index. php/Lebesgue_dimension.

[11] H.-O. Peitgen, P. H. Richter, *The Beauty of Fractals: Images of Complex Dynamical Systems*, Springer-Verlag, 1986.

[12] C. Eloy, "Leonardo's rule, self-similarity, and wind-induced stresses in trees," *Physical Review Letters,* vol. 107, 2011.

[13] L. Sallows, "On self-tiling tile sets," *Mathematics Magazine,* vol. 85, no. 5, pp. 323–333, December 2012.

[14] S. Northshield, *"Ford Circles and Spheres,"* Cornell University, 2015.

[15] D. Mumford, C. Series, D. Wright, *Indra's Pearls: The Vision of Felix Klein*, Cambridge University Press, 2002.

[16] C. T. McMullen, "The Mandelbrot set is universal," in *The Mandelbrot Set, Theme and Variations*, Cambridge: Cambridge University Press, 2000, pp. 1–18.

[17] D. White, "The Unravelling of the Real 3D Mandelbulb," November 2009. [Online]. Available: www.skytopia.com/project/fractal/mandelbulb.html. [Accessed 13 April 2020].

[18] R. Chen, "The Mandelbox Set," 1 April 2014. [Online]. Available: http://digitalfreepen.com/mandelbox370/. [Accessed 1 August 2019].

[19] M. Green, "The Buddhabrot Technique," 1993. [Online]. Available: http://superliminal.com/fractals/bbrot/bbrot.htm. [Accessed 13 April 2020].

[20] J. Baez, "The Beauty of Roots," December 2011. [Online]. Available: johncarlosbaez.wordpress.com/2011/12/11/the-beauty-of-roots/. [Accessed 13 April 2020].

[21] T. Lowe, "Fractal Automata," 2012. [Online]. Available: https://vimeo.com/36816056, https://vimeo.com/36951270, https://vimeo.com/36947960.

[22] H. Segur, *"Lecture 8: The Shallow-Water Equations,"* McGill University, 2009.

[23] M. T. Silva, "Ocean Surface Wave Spectrum," 2015. [Online]. Available: www.research-gate.net/publication/283722827_Ocean_Surface_Wave_Spectrum.

[24] I. R. Young, L. A. Verhagen, "The growth of fetch limited waves in water of finite depth. Part 2. Spectral evolution," *Coastal Engineering,* vol. 29, no. 1–2, pp. 77–99, 1996.

[25] R. Fossion *et al.*, "Scale invariance as a symmetry in physical and biological systems: listening to photons, bubbles and heartbeats," in *American Institute of Physics*, 2010.

[26] A. L. Goldberger *et al.*, "Nonlinear dynamics in sudden cardiac death syndrome: Heartrate oscillations and bifurcations," *Experientia,* vol. 44, pp. 983–987, 1988.

[27] "The Chaos Game," [Online]. Available: http://pi.math.cornell.edu/~lipa/mec/lesson4.html.

[28] M. L. Lapidus, "An overview of complex fractal dimensions: From fractal strings to fractal drums, and back," in *Horizons of Fractal Geometry and Complex Dimensions*, Contemporary Mathematics, vol. 731, *American Mathematical Society*, Providence, RI, 2019, pp. 143–245.

[29] M. L. Lapidus, M. Frankenhuijsen, *Fractal Geometry, Complex Dimensions and Zeta Functions: Geometry and Spectra of Fractal Strings*, second edition, Springer Monographs in Mathematics, Springer, New York, 2013.

[30] P. Straka, "Fractal Curvature Measures and Image Analysis," Arxiv, 2015.

[31] K. J. Falconer, *The Geometry of Fractal Sets*, Bristol: Cambridge University Press, vol. 85, 1985.

[32] Z. W. Zhu *et al.*, "On the lower bound of the Hausdorff measure of the Koch curve," *Acta Mathematica Sinica,* vol. 19, no. 4, pp. 715–728, 2003.

[33] P. Mora, "Estimate of the Hausdorff measure of the Sierpinski triangle," *Fractals,* vol. 17, no. 2, pp. 137–148, 2009.

[34] A. H. Reis, "Constructal view of scaling laws of river basins," *Geomorphology,* vol. 76, pp. 201–206, 2006.

[35] T. Lowe, "Improving image clarity using local feature dimension," *Image Processing,* vol. 9, no. 7, pp. 553–559, 2015.

[36] P. S. Wesson, "A new approach to scale invariant gravity," *Astronomy and Astrophysics,* vol. 119, pp. 145–152, 1982.

[37] A. V. Ramallo, "Introduction to the AdS/CFT correspondence," Arxiv, 2013.

[38] P. D. Mannheim, "Making the case for conformal gravity," *Foundations of Physics,* vol. 42, no. 3, pp. 338–420, 2012.

[39] B. Werness, "Voxel Automata Terrain," 2017. [Online]. Available: bitbucket.org/BWerness/voxel-automata-terrain/src/master/. [Accessed 13 April 2020].

[40] J. Rampe, "Softology's Blog," 2017. [Online]. Available: softologyblog.wordpress.com/2017/05/27/voxel-automata-terrain/. [Accessed 13 April 2020].

[41] B. B. Mandelbrot, "A primer of negative test dimensions and degrees of emptiness for latent sets," *Fractals,* vol. 17, no. 1, pp. 1–14, 2009.

[42] C. Tricot, "General Hausdorff functions, and the notion of one-sided measure and dimension," *Arkiv för Matematik,* vol. 48, no. 1, pp. 149–176, 2010.

[43] M. L. Lapidus, G. Radunovic, D. Zubrinic, *Fractal Zeta Functions and Fractal Drums: Higher-Dimensional Theory of Complex Dimensions*, Springer Monographs in Mathematics, Springer, New York, 2016.

Index